Trash *to* Cash

How Businesses Can
Save Money
and
Increase Profits

T0271287

Fran Berman

S^t_L

St. Lucie Press
Delray Beach, Florida

S$_{\text{L}}^{t}$

Published by
St. Lucie Press
100 E. Linton Blvd., Suite 403B
Delray Beach, FL 33483

This book is dedicated to a few key individuals whose influence was essential to my commitment to writing it:

To my husband, Joe Vidmar, who has risen above being annoyed at my inability to throw out anything without first trying to extend its use to us, give it away, or recycle it, and who has joined me on countless trips to the recycling centers and thrift stores and has supported me in many endeavors that have demanded my time and attention away from him.

To my parents, Ethel and Al, who, without realizing it, taught me how to make the most of what we have.

To the future generations of this planet without whom our lives have no long-term meaning.

TABLE OF CONTENTS

ACKNOWLEDGEMENTS xi

PART ONE: WHAT A MESS WE'RE IN!

CHAPTER 1—WHAT MESS? 3
Background 3
Where Does the Paper Go? 10
Where Can We Get More? 14

CHAPTER 2—HOW DID WE GET INTO THIS MESS? 21
Population Growth and Paper Consumption 21
Industry's Social Awakening 23
Information and the Paperless Society 23
Attitude 24

CHAPTER 3—WHY SHOULD WE GET OUT OF THIS MESS? .. 27
Decreasing Costs 27
Seeing is Believing 28
What Qualifies as Recyclable? 34
City of New York 40
Closing the Loop 42
Source Reduction 43

PART II: WHO'S GOTTEN OUT OF THIS MESS? CASE STUDIES

CHAPTER 4—AT&T 51
Background 51
Major Successes 51
Bedminster's Dynamic Duo 53
AT&T Power Systems, Mesquite, TX 71
AT&T, Chicago, IL 94

More AT&T Successes 98
Executive Perspective: Take It From the Top 102
Final Notes 103

CHAPTER 5—MCDONALD'S **107**
Waste Stream Challenges 107
Meeting the Challenge 109
Customers Expect It 110
The Best Way to Avoid Legislation is to Make It Unnecessary 112
It's Good Business 113
Maintaining Momentum 120
Personal Perspectives 121
Innovative Vision 125

CHAPTER 6—MERRILL LYNCH **131**
Background 131
CENYC 132
Across the River and Across the Country 137
Community Champions 140
Corporate and Personal Responsibility 142
Closing the Loop 145
Beyond Paper Recycling 148

CHAPTER 7—GIORDANO PAPER RECYCLING
CORPORATION **151**
Humble Beginnings 151
Timing Is Everything 152
He Who Laughs Last 154
Crystal Ball 154

CHAPTER 8—V. PONTE & SONS **157**
A Century of Service 157
Partnering for Success 157
Clearing the Hurdles 158
Up in Smoke 159
A Long-Term Investment 159

PART III: HOW DO WE GET OUT OF THIS MESS?

CHAPTER 9—WHAT IT'S WORTH? **163**
The Challenge 163

The Three "Rs" 163
Meeting the Challenge 164

CHAPTER 10—CORPORATE ACTION PLAN 167
Phase 1: Research and Planning—Where Are You? Where Are You
    Going? ....
    168
Phase 2: System Introduction and Involvement—How Are You
    Going to Get There? 177
Phase 3: System Implementation and Change—How Can You
    Ensure You Will Get There? 193
Phase 4: Evaluation, Adjustment, and Revitalization—Where Do
    We Go from Here? 194
Review the Process 196

PART IV: BEYOND THE PAPER MESS

CHAPTER 11—WHERE ELSE CAN WE GO FROM HERE? 201
Re-evaluate Office Supplies and Practices 201
Reduce Packaging Waste 201
Reduce Organic Waste... 202
Reduce Toxins 202
Reduce Vehicle Emissions.. 203
Reduce Energy Consumption 203
Reduce Water Consumption.. 203
Use Reusables 204
Donate Reusables 204
Recycle Plastics.. 204
Recycle Aluminum 205
Recycle Polystyrene 205
Redesign 206

CHAPTER 12—NEW BEGINNINGS 209

APPENDIX A 211

APPENDIX B 213

APPENDIX C 225

GLOSSARY 237

FOR FURTHER INFORMATION 239

INDEX 241

FIGURES AND TABLES

Figure 1-1.   Typical California municipal solid waste by weight 4
Figure 1-2.   1988 paper and paperboard consumption in the
              United States 11
Figure 2-1.   Worldwide population growth statistics 22
Figure 3-1.   Natural resources saved by manufacturing 1 ton of
              paper from recycled fibers. 29
Figure 3-2.   U.S. exports of recovered paper 33
Figure 3-3.   Production and re-production processes 35
Figure 3-4.   U.S. recovered paper consumption 1970–1993 37
Figure 3-5.   U.S. recovered paper utilization and collection 1970–
              1993 37
Figure 3-6.   U.S. total paper recovery rate 1970–1993 38
Figure 3-7.   Sample waste stream of office building with
              employee cafeteria 40
Figure 4-1.   AT&T total paper recycling 52
Figure 4-2.   AT&T paper use 52
Figure 4-3a. Recycling policy from AT&T employee handbook.... 58
Figure 4-3b. Waste minimization policy from AT&T employee
              handbook 59
Figure 4-4.   Janitor's note to noncomplying employee about
              trash 61
Figure 4-5.   Janitor's note to noncomplying employee about
              recyclables 62
Figure 4-6.   Cubic yards of trash sent to the landfill 74
Figure 4-7.   Source reduction equivalent in height, 1986–1991 .. 79
Figure 4-8.   Source reduction equivalent in miles, 1986-1991 80
Figure 4-9.   Data Center's paper reduction goal achievement 80
Figure 4-10. Comparison of printing costs by month 1991–1995 ... 81

Figure 4-11.   Daily average number of images printed from 1991
               to mid-1995 82
Figure 4-12.   Average number of pages printed per year from 1991
               to mid-1995 82
Figure 4-13.   Power Systems Division environmental awards won ... 88
Figure 4-14.   AT&T environmental, health and safety goals 102
Figure 4-15.   AT&T's environmental vision and policy 103
Figure 5-1.   Waste stream of a typical McDonald's restaurant ... 108
Figure 5-2.   McRecycle dollars spent 115
Figure 5-3.   McDonald's waste reduction policy 127–128
Figure 10-1.   Trash volume worksheet 168
Figure 10-2.   Disposal cost worksheet 168
Figure 10-3.   Waste audit worksheet 170
Figure 10-4.   Market development worksheet 171
Figure 10-5.   Monthly cost savings worksheet 172
Figure 10-6.   Waste reduction goal worksheet 173
Figure 10-7.   Eight steps guaranteed to reduce waste 197

Table 1-1.   Waste Reduction Goals in the United States 7–8
Table 1-2.   Examples of Recycled Content 9
Table 1-3.   Recommended Recycled Content 9
Table 1-4.   Some of the Largest and Smallest Consumer Nations
               of Paper and Paperboard 10
Table 1-5.   Quantities and Fate of Municipal Solid Waste 13
Table 1-6.   Annual Waste Generation in Selected OECD
               Countries, Late 1980s 14
Table 1-7.   Average Landfill Tipping Fees in the United States,
               1986–1990 14
Table 3-1.   Resources Saved by Using Waste Paper to Produce 1
               Ton of Recycled Paper 29
Table 3-2.   Common Recyclable Paper Products and Grade 35
Table 3-3.   Paper Products Reincarnated 36
Table 3-4.   Recycled Content in Paper and Paperboard 36
Table 3-5.   Recycling Requirements for Food or Beverage
               Service Establishments 41
Table 3-6.   Recycling Requirements for All Other Businesses 42
Table 4-1.   AT&T's Paper Products6 Recycling Results 53
Table 4-2.   Paper Recycling at AT&T–Bedminster, NJ 58
Table 4-3.   Paper Brands on Contract in 1995 66
Table 4-4.   AT&T-Dallas, TX: 1991 Recycling Volume 73

Table 4-5.   Cubic Yards of Trash Sent to the Landfill 73
Table 4-6.   Paper and Cardboard Rescued from Landfill in 1994 .. 74
Table 4-7.   Environmental Savings from Recycling 239 Tons of
              Paper and Cardboard 75
Table 4-8.   Comparison of Reimbursement Rates in the Dallas
              Area 1990 vs. 1995 75
Table 4-9.   Paper Recycling Revenues 1990–1995 75
Table 4-10.   Environmentally Friendly Products at AT&T Power
              Systems 87
Table 4-11.   Money Redeemed from Aluminum Cans for
              Donations 90
Table 4-12.   Waste Disposal Savings 1991–1994 93
Table 4-13.   AT&T Environmental Goals 99
Table 5-1.   U.S. Environmental Leadership Awards 110
Table 5-2.   McDonald's Environmental Partnerships 111
Table 5-3.   McDonald's Recycled Packaging 117
Table 5-4.   Waste Reduction Action Plan
              Accomplishments 118–119
Table 6-1.   Merrill Lynch Paper Recycling and Cost Avoidance
              Results for New York City Headquarters Facilities,
              1990–1994 137
Table 6-2.   Volume of Recycled Paper for Merrill Lynch, New
              Jersey Facilities 1990–1995 138
Table 6-3.   Natural Resources Saved 146
Table 6-4.   Merrill Lynch's Sources for Quality Recycled Paper .. 147
Table 10-1.   Approximate Waste Paper Reimbursement Rates
              in Chicago, IL 171
Table 11-1.   Forms of Some Recycled Plastics 205

ACKNOWLEDGMENTS

Dr. Sage Eileen Bennett was my instructor for the first course in my MBA program. She gave me an "A" on my project paper for her course on business ethics and social responsibility, and wrote "consider publishing this." *Trash to Cash* is based on that paper.

Many people have had a hand in guiding me on and editing this book. Several of them deserve special recognition, along with a number of others whose contributions do not directly appear in *Trash to Cash*.

Some of them are involved in promoting recycling, source reduction, and a sustainable economy: Ellen Lubell, Jennifer Pinkerton, Lynn France, Mia Finneman, Michael Dill, Suzy Moraes, and Yvette Berke.

Several were particularly helpful through the review and authorization process: Alison Pikus, Andy Lauro, Anna Steffanelli, Anthony Giordano, Sr., Bob Langert, Bobbie Collins, Carlee E. Weston, Jr., Cheryl LaPerna, George Perry, Jerry Twardy, Jim McMahon, Marilyn May, Mark Brownstein, Phil Lombard, Ron McCauley, Scott Van Buren, and Shelby Yastrow.

Others provided hard data that helped me translate statistics into terminology and images we can all relate to: Bill Buckalew, Mobil Chemical Consumer Products Div.; Bill Kelly and Marvin A. Crowson, South Coast Air Quality Management District; Craig Florer, Southern California Gas Co.; Robert S. Stone, California Environmental Protection Agency & Integrated Waste Management Board; Roy M. Matsuo, 76 Products Company; and Sandy Kravetz, Robert Marston & Associates, Inc.

My friends at NSA urged me to "write a book" and provided me the emotional support and technical expertise on how to do it, particularly Barbara Geraghty, George Morrissey, Jeff Slutsky, and Marilyn Snyder.

Several of my friends and family supported me by proofreading, researching, printing, or critiquing: Jo Ann Ferguson, Bob Allen, Trudy Grossman, Tricia LaRue, Linda Lewis, Raylene Rodriguez Brown, and Beth Ulrich.

Finally, my editor, Sandy Koskoff, shared her expertise and insight that will be invaluable in my future writings.

While I developed extensive expertise in writing, I benefited the most by realizing just how little I really knew about recycling and peoples' impact on the environment. I learned not just about paper recycling, rainforests, tree farms and lumber, but also about energy, plastics, glass, aluminum, and legislation.

Most of all, I had the privilege of talking and working with people and organizations dedicated to zero landfill and minimal environmental impact. If not for this book, I would never have known about them and about just how much is really being done to establish a sustainable society. I hope this book does them justice and serves them well in their environmental missions.

PART I:

WHAT A MESS
WE'RE IN!

CHAPTER I

WHAT MESS?

Background

What would our world be like without trees? The market for artificial Christmas trees would soar! Parasols might come back into fashion to provide shade. The plastics industry would take over the packaging market.

Besides for packaging, we'd have to find other materials to make into furniture and homes. Many parks already have benches and picnic tables made of heavy duty recycled plastic, and steel is growing in popularity within the residential construction industry for framing.

On a more serious note, our planet would be unlivable. Trees create oxygen that all living things need. They provide food and shelter for most of the animals alive today, which would otherwise become extinct. And what about their contribution to food and shelter for people—fruits, nuts, lumber?[1]

Clearly, trees are essential to our existence. But less than 5% of our old growth forests remain, and the population is growing faster than ever before. We are using up our trees, which are among our most precious natural resources, at an alarming rate.

Within the next 60 years, the global economy is projected to grow five-fold. In such a world, just holding the present, possibly unsustainable, environmental burden constant would require cutting the current environmental impact per unit of gross national product by 80 percent.[2]

At the same time, landfill closures and escalating trash disposal costs are forcing us to find alternative waste disposal methods and to change habits that exacerbate the situation. "Piled on a football field and tightly compacted, one year's worth of United States garbage—180 million tons in 1988—would reach a height of 26 miles."[3] In 1993, the United States generated enough garbage each year to fill a line of garbage trucks 145,000 miles long. That's more than halfway from here to the moon![4]

Californians alone generate 7 pounds of garbage per person per day, or about 46 million tons per year—enough to fill two freeway lanes of traffic 10 feet deep from Oregon to Mexico.[5] Figure 1-1 shows "Typical" municipal solid waste (MSW) in California consists of (by weight):[6]

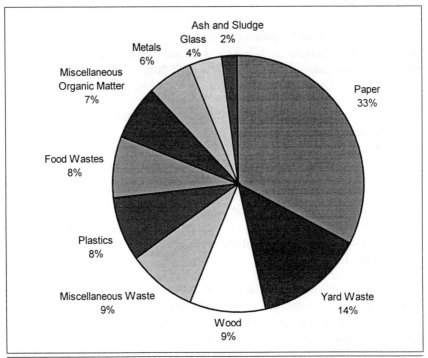

Figure 1-1. Typical California municipal solid waste by weight.

Government regulations require many cities and states across the country to recycle. This is also happening in other countries around the world. The fifteen member countries of the European Union have approved a directive "to put in place regulations to meet specified recovery and recycling targets. Each country must recover 50–60% of their packaging waste by mid-2001 and recycle at least 25–45%. At least 15% of all individual material types (glass, plastic, paper, etc.) must be recycled. The directive also requires the encouragement of recycled content in new products."[7]

In the United States, many states have passed legislation that mandates substantial waste reduction and/or recycling over the next several years, imposing heavy fines on cities and government agencies that don't comply.

According to the League of Women Voters Population Coalition, the U.S. population has grown from 150 million in 1950 to more than 250 million in 1994. Furthermore, the League foresees that figure could jump to 506 million by 2050.

The League recently reported that in California, population growth has skyrocketed from under 1.5 million in 1900 to over 32 million in 1995. The rate is expected to grow faster than the national average, and the garbage dilemma along with it. Consequently, its cities have no time to waste in getting residents, the public sector, and private industry on the band wagon.

To deal with all the garbage, California passed the Integrated Waste Management Act, AB939, in 1989. This legislation requires the diversion of 25% of the 1990 waste stream volume going into landfills by 1995 and 50% by 2001. One of the components includes waste reduction strategies. (It does not require 25% less generation, just 25% diversion.)

The city of New York mandates source separation by commercial firms. Local Law #19 states that a commercial building must recycle 50% of its waste stream. The waste haulers servicing those businesses must report the waste they pick up by categories and volume.

Dutchess County, about 100 miles north of New York City, reported a municipal solid waste volume of 5.1 pounds per capita for 1990. The projection for the year 2010 is 5.6 pounds per person, and a population increase of 14%. In other words, the amount of municipal solid waste would jump from 250,000 tons to 313,000 per year. The county's disposal costs would escalate by $9.45 million if rates remain at the 1990 figure of $150 per ton. If rates go up, and they already have, the cost to the county (and to individual and corporate taxpayers) goes even higher.[8]

According to U.S. EPA estimates, "In 1970 nondurable goods comprised 19 percent by weight of discards. This increased to 27 percent in 1990."[9] Fortunately, efforts to reduce, reuse, and recycle have become widespread since 1990 (see Table 1-1). But with a growing population and our throw-away habit, that's not enough to achieve admirable waste reduction goals.

On a national level, the federal government purchases over 300,000 tons of printing and writing paper annually.[10] President Clinton issued an Executive Order in the fall of 1993 mandating that all federal purchases of printing and writing paper contain a minimum of 50% recycled fiber with at least 20% post-consumer waste by the end of 1994, and 30% by the end of 1998. (California law requires that any consumer good labeled "recycled" must contain at least 10% post-consumer content by total product weight, shown in Table 1-2).

The Recycling Advisory Council of the National Recycling Coalition defines recycled paper and recommends recycled content (see Table 1-3) as follows:[11]

> Total Recycled Fiber...: Fiber derived from recovered paper excluding any paper generated in a paper mill prior to the completion of the paper manufacturing process.

> Post-consumer (or Comparable) Fiber...: Fiber derived from post-consumer materials as defined within RCRA Sec. 6002 and including fiber derived from recovered paper which has been printed and/or contains inks or colored dyes (excluding whitening or 'bluing' dyes or agents).

Similar actions are being taken outside the United Sates, too. In Barbados, waste paper has been going to the landfill at the rate of about 1,000 tons per week. The Barbados government recently decided to discontinue landfilling paper. No paper recycling plant on the island means baling and exporting waste paper, which can be very expensive.[12]

Table 1-1.   Waste Reduction Goals in the United States (baseline in parentheses)[13]

State	Source Reduction	Recycling	Waste Reduction
Alabama		25% by 1995	
Arkansas		30% by 1995 40% by 2000	
California			25% by 1995 50% by 2000 (1990 total waste)
Connecticut	No net increase in per capita waste 1990–2010	37% by 2010	
Delaware		30% by 1994	
District of Columbia		45% by 1994	
Florida		30% by 1994	
Georgia			25% by 1996 (1992 per capita)
Illinois			15% by 1994 25% by 1996 (1991 total waste)
Indiana			35% by 1996 50% by 2000 (1991 total waste)
Iowa			25% by 1994 50% by 2000 (1988 total waste)
Kentucky			25% by 2000 (1993 per capita)
Louisiana		25% by 1992	
Maine	5% by 1992 10% by 1994 (1990 total waste)	25% by 1992 50% by 1994	
Maryland		15-20% by 1992 20% by 1994 (state agencies)	
Massachusetts	10% by 2000 (1990 total waste)	46% by 2000	
Michigan	5–12% by 2005 (1989 total waste)		50% by 2005
Minnesota		25% greater MN 35% Twin Cities 40% state agencies by 1994	
Mississippi		25% by 1996	
Missouri		40% by 1998	
Nevada		25% by 1994	
New Hampshire			40% by 2000 (1990 per capita)

Table 1-1.   cont.

State	Source Reduction	Recycling	Waste Reduction
New Jersey	Cap waste gener-ation at 1990 baseline by 1996, reduce by 2000	60% by 1995	
New Mexico			25% by 1995 50% by 2000 (4 lbs. per day per capita)
New York	8-10% by 1997 (1987 per capita)	40–42% by 1997	
North Carolina			25% by 1993 40% by 2000 (1991 per capita)
North Dakota			10% by 1995 25% by 1997 40% by 2000 (1991 total waste)
Ohio			25% by 1995 (1989 total waste)
Oregon		50% by 2000 56% by 2006	
Pennsylvania	No increase generation from 1988 to 1997	25% by 1997	
Rhode Island		15% by 1994 residential	
South Carolina		25% by 1997	30% by 1997 (1993 total waste)
South Dakota			20% by 1995 35% by 2000 50% by 2005 (1990 total waste)
Tennessee			25% by 1995 (1989 per capita)
Texas		40% by 1994	
Vermont			40% by 2000 (1987 per capita)
Virginia		10% by 1992 15% by 1993 25% by 1995	
Washington			50% by 1995 (1990 total waste)
West Virginia		30% by 2000	

Government agencies, nonprofit organizations, and public educational institutions must deal with such regulations while operating costs escalate. Private industry has to fight shrinking margins in a highly competitive global market. Recycling, closing the loop by buying recycled

Table 1-2.   Examples of Recycled Content[14]

INDUSTRIAL WASTE CATEGORIES		
Manufacturers Manufacturing Waste Such as ...	**Converters Pre-Consumer Waste Such as ...**	**Consumers Post-Consumer Waste Such as ...**
•Discards from papermaking	•Misprinted paper	•Used office paper
•Scraps from steel and aluminum manufacturing	•Unsold magazines	•Discarded product packaging
•Bottles broken in the factory	•Cuttings from envelope production	•Used steel and aluminum cans
•Trimmings from plastic production	•Scrap from forming cans from steel and aluminum sheets	•Used glass bottles
		•Used plastic containers
	Pre-Consumer Waste	**Post-Consumer Waste**
	RECYCLED CONTENT CATEGORIES	

Table 1-3.   Recommended Recycled Content[15]

	Total Recycled Fiber %	Post-Consumer (or comparable) Fiber %
Newsprint:	40	40
Printing and Writing:		
Uncoated	50	15
Coated	40	10
Tissue:		
Bath, Towel, Napkins, Wipers	80	70
Facial	60	50
Construction Paper:	80	65
Paper Packaging:		
(Bleached & Unbleached)	40	20
Uncoated Paperboard:		
Corrugated Containers	40	35
Folding Cartons	100	60
Misc. Products (tags, tickets, etc.)	100	60
Industrial Paperboard	100	60
Uncoated Paperboard:		
Folding Cartons	90	45
Misc. Products	90	45

products, and source reduction (reducing consumption) are critical strategies in achieving positive outcomes in each of these areas.

As consumers and companies grow more and more concerned about Earth's deteriorating environment and the fate of future generations, they will insist on changes. Some of those changes include affordable renewable energy sources, environmentally friendly alternatives, and recycled goods. Communities, businesses, and nations are vying for limited natural resources. As constraints worsen, demand for recycled goods will rise. With increased demand and better remanufacturing technologies, goods with recycled content will be more competitively priced for the consumer. Recycling will become more profitable for everyone in the loop, from waste generator to re-manufacturer.

In particular, recycling paper products in the workplace poses significant advantages. It can enhance a company's public image and improve employee morale. More importantly, it can augment the company's bottom line by providing a source of incremental revenue and reducing expenses involved in waste disposal and paper consumption.

This book will examine the issues and challenges of developing a successful paper recycling program in the workplace. We'll see how some well-known companies have welcomed and met those challenges. Finally, easy to follow guidelines will set the stage for organizations of all sizes to succeed in similar programs.

Where Does the Paper Go?

Who uses all of our paper and paperboard? The greatest user in the world per person per year is the United States (see Table 1-4).

Table 1-4.   Some of the Largest and Smallest Consumer Nations of Paper and Paperboard[16]

	Pounds	Kilograms
United States	683	310
Finland	549	249
Japan	503	228
Belgium	470	213
Denmark	467	212
Canada	441	200
Sweden	441	200
Switzerland	441	200
Holland	441	200
Germany	441	200
United Kingdom	364	165
Italy	291	132
Brazil	57	26
Thailand	60	27
India	7	3

About 246 million tons of paper and paperboard are produced each year. Of that, over 154 million tons come from wood, more than 8 million tons from other natural fibers, and 95 million tons from reclaimed material.

Paper is classified into six main categories:[17]

- Newsprint

- Printing and writing papers

- Case-making materials

- Packaging papers and boards

- Household and toilet tissues

- Industrial and special purpose papers

In 1988, consumption of paper and paperboard in the U.S. reached 85.5 million tons. About 26 million tons of this were recovered for recycling purposes; 7.5 million tons went into permanent uses, such as books, permanent files and building construction; 8.4 million tons went into sewer systems or nonrecoverable uses, such as cigarette papers, home fireplaces

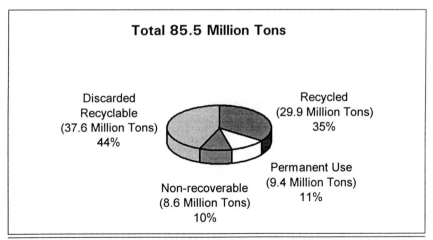

Total 85.5 Million Tons

Discarded
Recyclable
(37.6 Million Tons)
44%

Recycled
(29.9 Million Tons)
35%

Non-recoverable
(8.6 Million Tons)
10%

Permanent Use
(9.4 Million Tons)
11%

Figure 1-2.   1988 paper and paperboard consumption in the United States.

or were discarded in rural areas where collection systems do not exist. The remaining 43.6 million tons of paper and paperboard consumed in 1988 were discarded after use into municipal solid waste systems.[18]

By the end of 1993 "for the first time in U.S. history, more paper was recovered in the United States than was landfilled."[19] In 1994, "the U.S. recovery rate topped 40%."[20] Paper products make up 80% of all the recyclables recovered from the waste stream, and 85% of all paper manufactured is for business use.[21] That makes paper the most common and most available recyclable in corporate America today.

Where Will It All Go?

In the United States, "garbage generation on a per capita basis has soared to 4.3 pounds a day from 2.6 pounds per person per day just 30 years ago (see Table 1-5), and is projected by the U.S. Environmental Protection Agency to reach 4.9 pounds per person per day by 2010."[22]

Every day, U.S. streets carry more than 65,000 garbage trucks

> getting six miles per gallon of fossil fuels and belching carbon dioxide—a gas that is believed to be contributing significantly to global warming. Depending on where they are, anywhere from 45 percent to 60 percent of what these trucks pick up comes directly from offices, stores and factories.[23]

The amount of paper and paperboard in the U.S. waste stream is projected to grow to 121.2 million tons by 2010.[24]

"United States residents generate an average of 463 pounds of packaging waste per year, far more than residents of many other industrialized countries."[25] Packaging comprises one-third of the national waste stream by weight (see Table 1-6).

Landfills are quickly being exhausted in the United States and in other countries. Germany's

> Federal Environment Agency has estimated that of the country's 600 landfills, 500 will be closed by the year 2000 and that landfill capacity will be more or less exhausted 20

Table 1-5.   Quantities and Fate of Municipal Solid Waste (metric measure)[26]

Country	Quantities of MSW			Waste Disposal Routes			
	1980 (kg per inhabitant)	1990 (ktons per year)	Amount	Combustion	Landfill	Composting	Recycling
Austria	222	320	2,800	11	65	18	6
Belgium	313	343	3,500	54	43	0	3
Canada	524	601	16,000	8	80	2	10
France	260	328	20,000	42	45	10	3
Germany	348	318	25,000	36	46	2	16
Greece	259	296	3,150	0	100	0	0
Italy	249	348	17,500	16	74	7	3
Japan	355	408	50,000	75	20	5	N/A[27]
Luxembourg	351	448	180	75	22	1	2
The Netherlands	489	497	7,700	35	45	5	16
Norway	416	472	2,000	22	67	5	7
Portugal	214	287	2,650	0	85	15	0
Spain	270	322	13,300	6	65	17	13
Sweden	302	374	3,200	47	34	3	16
Switzerland	351	441	3,700	59	12	7	22
United Kingdom	319	398	30,000	8	90	0	2
United States	723	803	177,500	16	67	2	15

years later, unless waste volumes or public aversion to new sites can be drastically reduced.[28]

In the United States, there were over 14,000 landfills operating in 1978. That number dropped to 5,500 in 1988, and some project it to fall to 1,800 by 2000.[29] Two 100-year capacity landfills are planned for

Table 1-6.   Annual Waste Generation in Selected OECD[30] Countries, Late 1980s (pounds per capita)[31]

Country	Total Waste Generated	Amount of Packaging Waste
United States	920	463
Canada	1389	485
Finland	1120	289
The Netherlands	1033	344
Japan	876	359[32]
United Kingdom	793	295
Austria	789	291
Belgium	776	595
West Germany	707	276
France	673	399
Italy	669	242

Table 1-7.   Average Landfill Tipping Fees in the United States, 1986–1990 ($ per ton)[33]

Region	1986	1988	1990	1995 (June)[34]	1995 (July)[35]	Rate of Increase 1986–1990 (%)
Northeast	$17.57	$61.11	$64.79	$64.23	$63.77	269
Mid-Atlantic	$21.41	$33.84	$40.75	N/A	N/A	90
South	$11.86	$16.46	$16.92	$32.65	$32.50	43
Midwest	$11.75	$15.80	$23.02	$30.82	$31.18	96
West Central	$6.21	$10.63	$11.06	$20.53[36]	$19.70[37]	78
South Central	$8.71	$11.28	$12.50	N/A	N/A	44
West	$11.10	$19.45	$30.63	$35.07	$36.16	176

the California desert. However, escalating tipping fees (see Table 1-7) at current landfills signal the critical dilemma of having a sprawling, rapidly growing population generating more and more waste .

Exporting garbage over county and state lines, even to other countries, is not uncommon. However, this practice poses ethical issues about polluting untainted regions in this country and other countries, or even polluting outer space.

Where Can We Get More?

Perhaps the real question is "Can we get more?".

Forests and plant life are indispensable in producing oxygen and keeping our environment in balance. Entire old growth forests are being exterminated. Even with the most aggressive tree planting efforts, we can't readily replace the environmental support that older trees provide. We see this most graphically in the rapidly disappearing rainforests of the Amazon, of Southeast Asia, and of tropical third world countries like Madagascar. We see it closer to home, too, in Montana, Oregon, and other states where the lumber industry has flourished for decades. Most of these forests are cut down for construction, furniture, and export.

> The 460 million new pallets made in the USA annually contain half of all the hardwood timber cut down in the country each year. Overall, U.S. pallet makers consume 10% of all the lumber used in the country each year, and millions of the pallets bearing imported goods are made from tropical hardwoods.[38]

Many of these pallets are discarded in landfills rather than recycled. A large percentage of them is still in excellent condition and could be reused.

> Net annual deforestation during 1981–90 exceeded 2 percent in 10 tropical countries, all located in either Asia or the Americas. Brazil and Indonesia accounted for the largest extent of forest area lost annually, with deforestation levels of 3.7 million hectares and 1.2 million hectares on average per year, respectively.

> Annual logging of closed broadleaf forests during the 1980s averaged 5.6 million hectares, or 0.5% of the total broadleaf forest area. Eighty-four percent of this logging occurred in primary (undisturbed) forests.

> Forest loss in certain dry zones is also of concern: in Asia, almost 3% of forests in very dry areas were lost annually while deforestation rates in the Americas were highest in desert areas, averaging 2% during 1981–90.

> Global roundwood production increased 19% between 1979–81 and 1989–91. This increase was highest in Africa, where production rose by one third during this period. Most of the roundwood cut in Africa (about 90%) is used for fuelwood and charcoal, as opposed to industrial purposes such as for construction or paper production. Globally, about half of the

roundwood produced in 1989–91 was used for heating and cooking.[39]

Trees manufactured into paper products come primarily from tree farms, so paper recycling doesn't really do much to save old growth trees and tropical rainforests. But if we can make paper from other paper, we can use the farmed trees for lumber and safeguard our old growth trees for our children, our grandchildren, and their grandchildren. At the rate of current trends, we will leave them a world poor in natural resources.

"A single forest tree absorbs 26 pounds of carbon dioxide per year. An acre of trees can remove 2.4 to 5 tons of carbon dioxide per year."[40] Without the cleansing benefits of the forests, serious illnesses will abound, their severity will increase with environmental degradation, and our families will have few options.

Here Today, Gone Tomorrow

To understand the potential impact on our society we can look at the history of Easter Island in the South Pacific. "When Europeans arrived, the native animals included nothing larger than insects–not a single species of bat, land snail, or lizard."[41] Researchers are in the process of conducting the first systematic excavations on Easter Island and have come to some startling conclusions. "For at least 30,000 years before human arrival and during the early years of Polynesian settlement, Easter was not a wasteland at all."[42] It had flourishing subtropical forests and plants, a wide variety of flora, fauna, fowl, animals, and fish, most of which are now extinct. The research team has concluded that the early Polynesians enjoyed this rich landscape. After a few centuries, however, greed led to overexploitation of the island's natural resources. "Eventually Easter's population was cutting the forest more rapidly than the forest was regenerating. The people used the land for gardens and the wood for fuel, canoes, and houses—and, of course, for lugging statues. As forest disappeared, the islanders ran out of timber and rope to transport and erect their statues. Life became more uncomfortable–springs and streams dried up, and wood was no longer available for fires....Land birds, large sea snails, and many seabirds disappeared....Fish catches declined..."[43] as did crop yields "since deforestation allowed the soil to be eroded by rain and wind, dried by the sun, and its nutrients to be leeched from it."[44] By the time European explorers landed in 1722, Easter Island was practically barren and its inhabitants were in a frail and pitiable state. Can it be that we are going down the same path as the early civilization on Easter Island? Perhaps.

"Recycling half the paper used throughout the world today would free 20 million acres of forest land from paper production."[45] Yet the world's rainforests are being destroyed for logging, mining, and cattle grazing at the rate of 50 acres per minute, or "27 million acres per year— (an area) roughly the size of Pennsylvania. According to the Rainforest Action Network, two-fifths of the rainforests are gone already, and at the current pace, total decimation would occur in the year 2057."[46]

Here are some other startling statistics. "At the rate we're using up bauxite, the Earth will be completely stripped of it in 200–300 years...The known oil reserves in the world will last only an estimated 35 years at the rate we're using them[47]...The U.S. loses enough [top]soil every year to fill 50 million boxcars."[48]

While the statistics do not paint a very pretty picture, the picture is by no means complete. There is still time.

ENDNOTES

1. *The Day the Trees Disappeared*, COMIX CCA.
2. *World Resources 1994-95* (New York, NY: Basic Books), p. 217.
3. *INFORM Annual Report 1992: Strategies for a Better Environment* (New York, NY: INFORM, Inc., 1992), p. 7.
4. *50 Simple Things Kids Can Do to Recycle* (Berkeley, CA: EarthWorks Press, Inc., and California Department of Conservation, 1994), p. 10.
5. *Educator's Waste Management Resource & Activity Guide* (Sacramento, CA: California Department of Conservation, Division of Recycling, April 1992), p. 81.
6. Ibid.
7. *The Warmer Bulletin, Number 46* (Tonbridge, Kent, UK: The World Resource Foundation, August 1995), p. 3.
8. Dutchess County Draft Generic Environmental Impact Statement and Solid Waste Management Plan, March 1991 (cited in Bette K. Fishbein and Caroline Gelb, *Making Less Garbage: A Planning Guide for Communities*), pp. 36–37.
9. *Closing the Loop: Integrated Waste Management Activities for School and Home, K-12 edition* (Chagrin Falls, OH: The Institute for Environmental Education, 1993), D-44.
10. *The Recycling Advocate* (Sacramento, CA: Californians Against Waste Foundation, December 1993), p. 2.
11. Recycling Advisory Council, *Fact Sheet: Standards & Definitions for Recycled Paper* (Washington, DC: National Recycling Coalition, June 1992), p. 1.

12. *The Warmer Bulletin, Number 43* (Tonbridge, Kent, UK: The World Resource Foundation) November 1994, p. 5.

13. Bette K. Fishbein and Caroline Gelb, *Making Less Garbage: A Planning Guide for Communities* (New York, NY: INFORM, Inc., 1992), pp. 26–29.

14. Recycled Paper Coalition of California, 1994.

15. Recycling Advisory Council, *Fact Sheet: Standards & Definitions for Recycled Paper* (Washington, DC: National Recycling Coalition, June 1992).

16. *The Warmer Bulletin, Number 43, Information Sheet: Paper Making & Recycling*, p. 2.

17. Ibid., p. 1.

18. *12 Facts About Waste Paper Recycling* (New York, NY, NY: American Paper Institute, Inc., 1989), section 8.

19. *PaperMatcher: A Directory of Paper Recycling Resources* (Washington, DC: American Forest & Paper Association, October 1994), p. i.

20. *Recovered Paper Statistical Highlights: 1995 Edition* (Washington, DC: American Forest & Paper Association, July 1995), p. 1.

21. The United States Conference of Mayors, *The National Office Paper Recycling Strategy* (Washington, DC: National Office Paper Recycling, 1991), p. 2.

22. *INFORM Annual Report 1992* (New York, NY: INFORM, Inc., 1992), p. 7.

23. Lee Wessman, *Business and the Earth: Global Ideas for Local Solutions* (CityBusiness/USA, Inc., 1991), p. IV.

24. *Characterization of Municipal Solid Waste in the United States: 1990 Update* (Washington, DC: U.S. EPA, June 1990), pp. 10, 59 (cited in B. Fishbein and C. Gelb, 1992, p. 8).

25. B. Fishbein and C. Gelb, p. 116.

26. *The Warmer Bulletin, Number 44: Information Sheet* (Tonbridge, Kent, UK: The World Resource Foundation, February 1995), p. 3.

27. MSW levels in Japan are calculated after the removal of recyclables.

28. *The Warmer Bulletin, Number 43*, p. 9.

29. Marketing News, *Economics Meets Ecology as Recycled Paper Matures* (Chicago, IL: American Marketing Association, 1994).

30. Organization for Economic Cooperation Development.

31. James E. McCarthy, *Recycling and Reducing Waste: How the United States Compares to Other Countries* (Washington, DC: Congressional Research Service Report for Congress, November 8, 1991), p. 8 (cited in B. Fishbein and C. Gelb, pp. 7, 116).

32. Represents consumption of packaging, not waste; therefore, waste data would be substantially lower.

33. *National Solid Waste Management Association Survey Tracks Upward Trend in Landfill Tipping Fees* (Washington, DC: Integrated Waste Management, December 11, 1991), p. 7 (cited in B. Fishbein and C. Gelb, p. 16).

34. *Solid Waste Digest: Vol. 5, Number 6* (Alexandria, VA: Chartwell Publications, June 1995), p. i.

35. *Solid Waste Digest, Vol. 5, Number 7* (Alexandria, VA: Chartwell Publications, July 1995), p. i.

36. Solid Waste Digest defines this as the West region, excluding the West Coast which it defines as the Pacific region.

37. Ibid.

38. *The Warmer Bulletin, Number 43*, p. 18.

39. *World Resources 1994-95* (New York, NY: Basic Books), p. 305.

40. *Educator's Waste Management Resource & Activity Guide*, p. 3.

41. Jared Diamond, "Easter's End", *Discover Magazine* (August 1995), p. 64.

42. Ibid., p. 66.

43. Ibid.

44. Ibid.

45. *Closing the Loop: Integrated Waste Management Activities for School and Home, K-12 edition*, D-44.

46. Lee Wessman, *Business and the Earth: Global Ideas for Local Solutions*, p. V.

47. *15 Simple Things Californians Can Do to Recycle* (Berkeley, CA: EarthWorks Press, Inc., 1991), p. 10.

48. Ibid., p. 13.

CHAPTER 2

How Did We Get Into This Mess?

There are many factors that have brought about these conditions. The three main ones are:

(1) a rapidly increasing population,

(2) unchecked consumption, and

(3) the advent of the information age.

Population Growth and Paper Consumption

At the 1994 Cairo Conference on Population, it became strikingly clear that the extensive consumption by the industrialized world demands more and more of our air, land, and water resources every year. As the developing world strives to achieve similar life styles and standards of living, those demands are becoming even more critical. If the largest consuming nations like the United States don't set a good example, how can we dictate to developing countries that they conserve or do without?

The world's population has catapulted from 1 billion in 1830 to 5.6 billion in 1995, and it is expected to double to 11 billion by the year 2030.[1] Without any change in current conditions and habits, consumption automatically rises in direct proportion to the population.

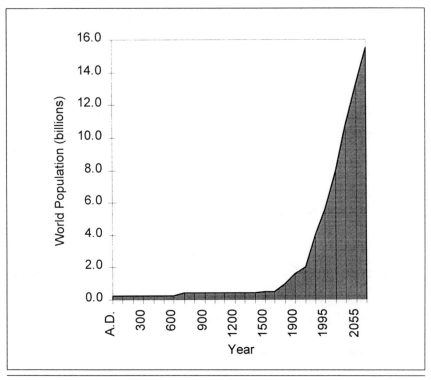

Figure 2-1.   Worldwide population growth statistics (projected to 2080).

The 1.1 billion people who currently live in industrialized countries represent only 20% of the 5.6 billion total. Yet they also represent:

80% of the world's resources consumed

85% of the consumption of the world's forest products

72% of all steel products consumed

75% of all energy consumed globally

75% of the global burden of pollutants and waste[2]

"United States' population is growing by 2 million people a year—not counting immigration (legal and illegal); each U.S. citizen consumes 40 times more natural resources than a citizen from a poor country."[3]

Industry's Social Awakening

In less than a century, industry has introduced new machines, technologies, and convenience tools that have made record profits for their companies while inflicting often irreversible harm on the environment. While some damage has been done knowingly and with little conscience, much of it was the unintentional result of short-sightedness, greed, and lack of concern for the long-term effects.

Companies, concerned only with short term profits and not long term costs to society, did not take responsibility for their actions. Consumers purchased and discarded without looking beyond their own turf, earning the nickname "the throw-away society."

Fortunately, all that is changing, as it must if we are to reap the intangible profits the world has to offer.

Information and the Paperless Society

Office automation brought with it the vision of the "paperless office," a vision that information would be created, transmitted, shared, and stored electronically. This has taken place to a large degree. Unfortunately, however, we often find ourselves generating more paper, not less. Computer users print drafts, reports, and hard copies of files without a second thought to the resources being used. Computerized organizations often find themselves with more paper than ever.

It's surprising what a small percentage of the available computing power is really used in computerized companies, even within the computer industry itself. This is partly because management doesn't insist on maximizing the capabilities of their computer investments. They encourage, and even oppose, training employees on how to use the software on their PCs. Frequently, managers themselves barely scratch the surface of the technology on their own desks. Many managers don't even have their own computers, preferring to rely on their skilled support staffs. Still others may print (or have their secretaries print) their electronic mail messages. They either don't have time to read their e-mail on the computer screen, can't (or won't) wean themselves from the hard copy, or don't know how to use the software (and won't let anyone know).

As givers of information, we feel compelled to give everyone a copy, traditionally single-sided. As receivers of information, we're addicted to having our own copy, even if we never refer to it again. In 1990, over 5 million photocopiers in the United States made close to 500 billion copies using almost 2 million tons of paper.[4] Imagine how much time we spent filing all that paper!

Attitude

Unfortunately, most companies and private citizens have not taken paper recycling seriously. Why not? Companies and individuals have similar reasons.

Objection: "It takes too much time, and time is money."

Response: How much money is wasted on excessive layers of packaging when simpler packaging will suffice? How much money is thrown away on paper towels where sponges or cloth towels could be used and reused? What about all that "junk mail" that ends up in the "circular file" instead of being recycled, bringing incremental revenue to the waste generator? "According to a study completed for the Environmental Protection Agency, the total amount of time a householder spends recycling is about 73 minutes per month. That's a little more than 2 minutes a day."[5]

Objection: "The recycle bins take up too much space" or "are in the way."

Response: How much space is discarded paper taking up in our landfills? Since 85% of business waste is paper, recycle bins are merely replacing existing trash containers. Containers for the remaining waste will be considerably smaller.

Objection: "It's not convenient."

Response: Does it have to be as convenient as pressing the remote control? As the case studies in this book illustrate, the most successful recycling programs are the ones that are the easiest to set up and the easiest to follow. If mammoth companies like AT&T, McDonald's, and Merrill Lynch can do it with thousands of employees, so can the rest of us.

Objection: "There's an overabundance of recyclables waiting to be re-manufactured, so why add to this excess supply?"

Response: The solution is a closed loop. Consumers must demand goods made of recycled material, otherwise we're just trading one type of garbage pile for another. Increased demand for recycled goods, improved remanufacturing techniques, and higher costs for virgin fiber will support attractive reimbursement rates for recycled fiber.

Objection: "Paper is made from natural fibers, so it will biodegrade anyway."

Response: For paper to biodegrade, it must have air. Landfills are too tightly packed to let enough air in for the natural process to occur. Besides, although many of the trees for paper production are grown as a cash crop on tree farms, we're using them much faster than we're replacing them.

When management sets the standard, employees will become more aware of waste in the workplace. Together they can work to change habits and make a positive impact on both the organization's and Earth's bottom lines.

ENDNOTES

1.   David Saphire, *Making Less Garbage on Campus: A Hands-On Guide* (New York, NY: INFORM, Inc., 1995), p. 5.

2.   Ibid.

3.   *We Need You to Help Curb Population Growth* (Santa Monica, CA: Population Education Committee, 1992).

4.   *Reducing Office Paper Waste: An INFORM Special Report* (New York, NY: INFORM, Inc., 1991), p. 7.

5.   On-hold message, Giordano Paper Recycling Corporation, Newark, NJ, 1995.

CHAPTER 3

WHY SHOULD WE GET OUT OF THIS MESS?

Why recycle paper? Aren't we supporting industry and jobs by consuming paper products? Don't we have plenty of trees? Can't we plant more?

Decreasing Costs

Worldwide, paper products are the second largest component of municipal solid waste:[1]

31% soil	6% metal
30% paper	6% textiles
12% plastic	8% other
7% yard waste	

The United States produced over 71 million tons of paper in 1991, 25% more than a decade earlier.[2] Of that, 6.6 million tons were exported.[3] With extensive recycling campaigns, the amount of recycled paper exceeded the volume that was landfilled for the first time ever in 1993.[4] By the end of 1994, the recovery rate had risen to 40.3%, or just over 38 million of the approximately 96 million tons of paper and paperboard produced.[5] Still, that means that the United States threw away over 57 million tons. That's enough paper annually to build a stack of paper as high as a two-story house for every man, woman, and child in the country.[6] It's enough to fill more than three pickup-trucks full for an American family of four every year. Every year, we throw out enough white writing paper to build a 12-foot-high wall of paper...from New York to Los Angeles."[7]

It's easy to see how recycling in the United States, with a population of over 260 million, lightens the burden on municipal solid waste collection and on waste disposal systems. It saves valuable space in landfills and extends their lives. This in turn saves the waste management industry the costs of developing new landfill sites, of building waste-to-energy plants, and of exporting garbage to other states and countries. Without recycling, waste disposal costs would continue to rise and ultimately be passed on to the consumer. We have the choice of paying now or paying later. If we pay later, the "price" will certainly be a lot more, and costs will not necessarily be measured in monetary terms.

Seeing is Believing

Let's put all this in perspective.

"Making one ton of paper requires nearly 3,700 pounds of wood, over 200 pounds of lime, 360 pounds of salt cake, 76 pounds of soda ash, 24,000 gallons of water, and 28 million BTUs of energy. In addition, making paper from raw materials means we must treat and dispose of 84 pounds of air pollutants, 36 pounds of water pollutants, and 176 pounds of solid waste."[8]

Recycling paper can be very economical in non-monetary terms. Manufacturing paper from recycled material instead of virgin fiber saves:[9,10]

Water	nearly 60%
Energy consumption	40%
Air pollution	74%
Water pollution	35%

But what does 1 ton of paper really look like?

Table 3-1.   Resources Saved by Using Waste Paper to Produce 1 Ton of Recycled Paper

Recycling one ton of paper saves...	Enough to...
17 oxygen producing trees[11]	produce 7,000 copies of a national newspaper[12]
2 barrels of oil (84 gallons)	propel the average American car 1,260 miles Dallas to Los Angeles[13]
7,000 gallons of water	overflow the average hot tub or spa 18 times
4,100 kilowatts of electricity	power the average home for six months[14]
53 million BTUs	run a gas clothes dryer for 331.25 hours, or 2 weeks non-stop day and night[15]
3.2 cubic yards of landfill space	fill one family size pick-up truck
60 pounds of air pollution (a 70% reduction)	eliminate the amount of nitrogen oxides that an average late model American car would emit driving 19,427 miles[16]

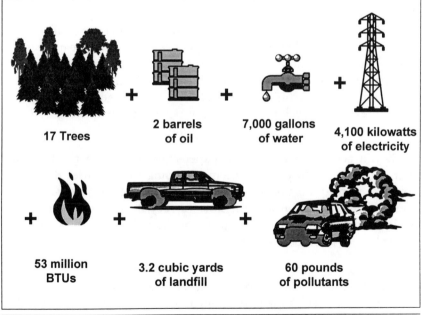

Figure 3-1.   Natural resources saved by manufacturing 1 ton of paper from recycled fibers.

When we combine these figures with the quantity of paper the United States discards every year—close to 50 million tons—we get some staggering statistics.

Seeing the Forest for the Trees

If recycling just 1 ton of paper saves 17 trees, then recycling 50 million tons saves 850 million trees—a forest almost the size of Connecticut.[17]

Newspapers are the third largest component of waste paper. Every day, people in the United States buy over 62 million newspapers. Unfortunately, almost 70% of them—about 44 million newspapers—end up in landfills each day. That's the equivalent of over 500,000 trees each week, or about 26 million trees per year.[18] At 300 trees per acre, it's like clear-cutting almost 87,000 acres every year. That's equivalent to 135 square miles, more than twice the size of Washington, DC. At that rate, a forested area the size of Rhode Island would be a treeless wasteland in less than eight years.[19]

Recycling one stack of newspapers about 6 feet-high saves one tree 35-feet tall.[20] "Recycling a 36-inch tall stack of newspaper saves the equivalent of about 14% of the average household electric bill."[21] One person recycling his/her daily newspaper every day of the year saves five trees.[22] A subscription to a major metropolitan newspaper for daily delivery consumes large amounts of trees and other resources when projected over a person's working life (44 years), ages 21 through 65. Just one person recycling his/her daily newspaper religiously can save 267 acres of trees, almost 0.5 square mile.

If you actually read the newspaper every day, then the trees and energy consumed to produce them are going to good use. But if you don't get to read most of those daily issues, change your daily subscription to Sunday only. And remember to recycle it, too. "If Americans recycled their newspapers for just one Sunday, we could save the equivalent of 550,000 trees!"[23]

Every day, mailrooms across the country deliver hundreds of thousands of trade journals and business magazines to employees who don't get to read them. Besides high environmental costs, these carry high labor costs. Cancelling unnecessary subscriptions and sharing publications can lower mailroom volume, employee costs, and paper consumption.

Water, Water Everywhere

How much water does it take to make 50 million tons of paper goods? The average above-ground spa or hot tub holds about 350 to 400 gallons of water. Recycling 1 ton of paper from virgin fiber saves 7,000 gallons of water, enough to overflow the average spa or hot tub more than 18 times. Recycling 50 million tons of paper saves about the same amount of water as Niagara Falls gushes[24] in a little over two-and-one-half days.[25]

> Paper making using virgin pulp requires as much as 300 cubic metres of water for every tonne of paper produced, while producing recycled paper from waste paper needs only 30 cubic metres of water. Even that reduced consumption can be further reduced by careful husbanding of resources, with such initiatives as recovering the steam used to dry new paper.[26]

Energize

Energy consumption, prices, and expenditures have been rising dramatically since the early 1970s. Although energy consumption by the paper industry is expected to grow only modestly over the next fifteen years, that growth could be lower if the paper recycling rate increases significantly. Recycling 1 ton of paper saves roughly 40%[27] of the energy needed to make it from trees.

As stated earlier, two barrels of oil per ton is enough to drive the average American car 1,260 miles, from Los Angeles to Dallas. Recycling 50 million tons of paper saves 100 million barrels of oil. With that, 83,333 people (one person per car) could make that trip.

Let's look at another energy resource: electricity. Recycling 1 ton of paper saves 4,100 kilowatts of electricity per ton, enough to heat and cool the average American home for six months.[28] Recycling 50 million tons of paper would save 205 billion kilowatts per year—enough electricity to power every American home for 14 years.

The Breath of Life

What about the air we breathe? Manufacturing 1 ton of recycled paper saves 60 pounds of air pollution. Recycling 50 million tons of paper would prevent the air pollution comparable to driving 50 million average late model American cars coast to coast four times. And then we must consider the pollution from generating the electricity and refining the oil. Employee absences and health expenses related to air pollution are very costly. This nation's "pollution abatement costs are rising faster than the rate of industrial production...U.S. manufacturers spend more than $40 billion annually on pollution control."[29] Manufacturing paper with recycled fiber and closing the loop by buying recycled paper, then, can cut down on air pollution and related illnesses by using less energy.

Using recycled paper to manufacture 1 ton of paper goods saves 3.2 cubic yards of landfill space. This is equal to filling a family size pick-up truck. It also saves money for businesses, residents, and taxpayers. Average disposal fees can range from under $20 per ton in some California counties to over $155 per ton in the New York City area where available land for new landfills is scarce or expensive.[30]

On the brighter side, recycling paper products is a growing industry that creates new opportunities and jobs.

> Incineration does not spur development of new business opportunities or directly create many jobs. Recycling, in contrast, uses solid waste as a material to create new products. It therefore has the potential to establish a new base for economic development in the City [of New York]....Recycling can create up to four times as many local jobs as incineration... Important for employment stability, recycling creates more permanent jobs in the operations and maintenance. Many of those jobs are entry-level low-skilled positions which can be filled by economically disadvantaged citizens.[31]

"For every 10,000 tons of waste materials recycled, 32.6 jobs are supported, compared to only 6.46 jobs supported when this much waste is landfilled."[32] According to a report by the Californians Against Waste Foundation (CAWF), Californians recycle over 2.6 million tons of paper every year, sustaining about 9,000 jobs.[33] The CAWF report estimated a total number of approximately 18,000 jobs supported by all types of recycling in a variety of areas:[34] 219 recycling-based manufacturers and end users; 2,844 recyclers, collectors, and processing facilities; and 465 companies operating curbside programs.

Furthermore, the "California Integrated Waste Management Board estimates that meeting the state's goal of 50% waste reduction and recycling could create 45,000 new jobs" by the year 2000. "About half of these will be new manufacturing jobs, which generally pay better than collection or processing jobs."[35]

"Not only has commercial solid waste disposal been reduced by 24%, but 80% of the businesses report that they saved money or incurred no additional cost to recycle."[36]

Paper recycling provides raw materials to much of the paper industry. At one time, it was a significant source of income for the recycler whose mainstay was recycling glass, aluminum, and metals. The market has developed so well that some companies make their livings by recycling paper alone.

Paper recycling also fuels the paper exporting business. The United States exported 408,000 tons in 1970. That grew to 6.6 million tons in 1991.[37] In 1993, recovered paper export revenues totaled $560 million[38] compared to $26 million spent on imports.[39] With international agreements such as NAFTA opening foreign markets, these figures should continue their upward trend.

"Shipments of (all) paper and allied products are expected to grow at an average annual rate of more than 2% through 1998."[40] Paper exports totaled $9.8 billion in 1993, accounting for close to 8% of all U.S. paper goods shipped.[41] Exporting recovered paper also helps offset the U.S. foreign trade deficit (Figure 3-2).

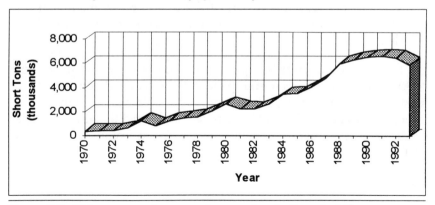

Figure 3-2. U.S. exports of recovered paper (total all grades).[42]

As international markets continue to grow, they must also foster widespread paper recycling. Most of the trees used for paper products come from tree farms. However, a significant increase in market growth will deplete the acreage for other plants, animals, and uses. Our environment could come under attack far more significantly than we can imagine.

What Qualifies as Recyclable?

The most common office paper is "dual purpose" for use in copiers and laser printers. Most office paper manufacturers make at least one type of recycled dual purpose office paper that is usually 10 to 30% post-consumer and 10 to 50% pre-consumer fiber. The term "post-consumer" applies to paper that has been purchased by the end user. "Pre-consumer" or "post-industrial" refers to everything that does not reach the consumer after the initial manufacture. That includes leftovers after milling, such as scraps on the mill floors and finished goods that were returned to the manufacturer unsold by distributors and retailers.

Paper products that are acceptable for recycling include: "dual purpose" desktop white paper commonly used in copiers and laser printers; corrugated boxes; newspaper; magazines; scraps and excess from paper manufacturing facilities; and "mixed office waste." Computer "greenbar" can also be recycled and commands the highest price of all recyclable paper products.

The following three charts provide more information on recyclability. Table 3-2 lists common recyclable paper products.[43] Figure 3-3 illustrates how paper recycling is part of the overall paper manufacturing process. Table 3-3 shows what form these paper products take when recycled.

The American Forest & Paper Association projects that in the year 2000, 78% of all U.S. recovered paper will be recycled domestically to make new paper and paperboard products, 15% will be exported to foreign recyclers, and 7% will be reused to make traditionally non-paper products such as animal bedding, insulation, hydromulch, and compost.[44]

Table 3-2.   Common Recyclable Paper Products and Grade[45]

Corrugated, old boxes
Corrugated, new cuttings
News, printed (old and overissue)
News, unprinted
Other groundwood papers (telephone directories)
Boxboard cuttings
Mixed (glossy, magazines, catalogs, windowed envelopes, sticky notes)
Brown kraft
Bleached kraft
Printed bleached kraft
Ledger Grades (computer paper, copy paper, letterhead, white notebook paper)
Old Corrugated Cartons
Other (wet strength with polycoatings, plastic, foil or hot melt glues; carbon paper;
    books with covers; colored   tabulating cards; carbonless forms; file
    stock; plastic-windowed envelopes; textile boxes; paper stock not
    otherwise listed)

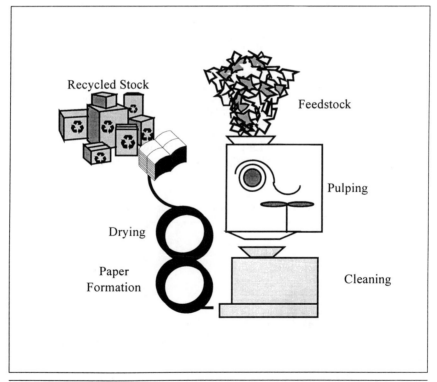

Figure 3-3.   Production and re-production processes.[46]

Table 3-3.   Paper Products Reincarnated

These...	can be recycled into products such as...
Cellulose Insulation	Office construction drywall
Computer Paper	Carbonless paper; continuous bond; form bond; computer greenbar
Office Supplies	Adding machine rolls; binders; dividers; files; folders; report covers
Packaging Materials	Boxes; cushioning; kraft envelopes; mailing tubes; molded packaging for eggs, fruit, vegetables, and other fragile items
Paper Products	Absorbents; paper refuse bags; books and journals; calendars; coloring books; file boxes; office recycling containers; food service containers (plates, bowls, trays, etc.)
Office Papers	Linen pads; loose leaf; note pads; spiral bound notebooks; telephone message pads; wrapping paper
Paperboard	Indexes; hanging files; kraft files; liner board; corrugating medium; pressboard; tube stock
Printing Papers	Bond; book; coated offset; copy and xerographic; cotton fiber; cover stock; envelopes; business cards; labels; mimeo; newsprint; offset; text paper
Tissue Papers	Industrial wipers; napkins; bath tissue; facial tissue; paper towels
Shredded Paper	Protective packaging (instead of polystyrene peanuts); cat litter; animal bedding
Other Uses	Plasterboard for home construction; thermal insulation; cat litter; molded disposable hospital products

The United States has a long way to go to catch up with the recycled paper content that some of the other industrialized nations have achieved (Table 3-4).

Table 3-4.   Recycled Content in Paper and Paperboard[47]

Denmark	70%
Spain	60%
United Kingdom	55%

But we are making significant progress in areas like newsprint. Newsprint can come back as newsprint up to three times. After that, unless virgin fibers are added, there's too little fiber left, and the resulting product tears easily in the printing press. Paper mills add scraps, sawdust, excess, and new virgin fiber to the recycled material to produce an acceptable grade of newsprint. Figures 3-4, 3-5, and 3-6 show the growth in recovered paper consumption and utilization from 1970 through 1993.

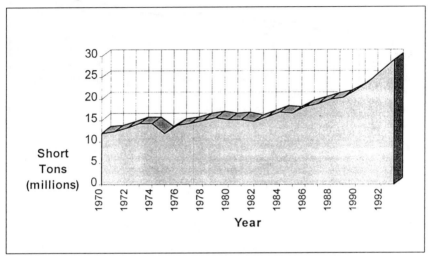

Figure 3-4.   U.S. recovered paper consumption 1970–1993.[48]

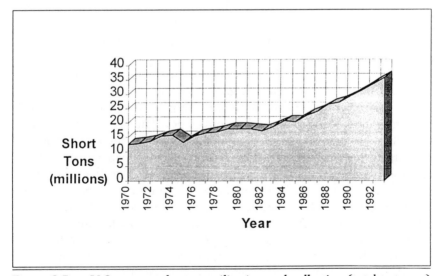

Figure 3-5.   U.S. recovered paper utilization and collection (total recovery) 1970–1993[49]

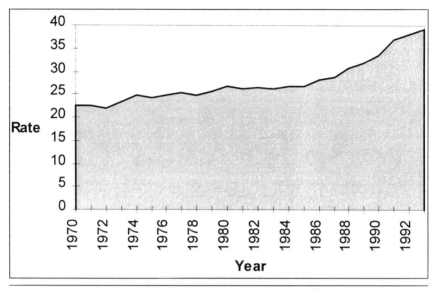

Figure 3-6.   U.S. total paper recovery rate 1970–1993[50]

Some mills produce 100% recycled "groundwood" paper, which can also be recycled. However, there is a lot of controversy about the use of groundwood paper among paper manufacturers and recyclers. Groundwood pulp is derived from wood fiber via mechanical pulping processes and uses the entire tree, including bark. Chemically processing paper does not use the bark, wasting a fair percentage of the tree. Groundwood paper fibers are shorter so that the finished product is not as strong—a concern for use in copiers and printers. Groundwood paper is slightly gray in color, not the bright white we're used to seeing. In its favor, it is more opaque than chemically processed paper, making it ideal for two-sided copying and for envelopes. The major issues have to do with (1) cosmetics/aesthetics, and (2) the resources used in the manufacturing process. Traditional desktop white paper is in demand and therefore commands a higher reimbursement rate than the grayer groundwood paper. A factor in favor of groundwood paper is that processing requires few chemicals and produces very little waste. The production of groundwood office paper uses a much wider range of recyclable materials such as newspapers, magazines, and catalogs. On the other hand, processing the virgin groundwood fiber uses more energy than its bright white counterpart, although this is not true in the recycling process. There is obviously more research to be done to establish an environmentally substantiated position on groundwood papers.

Recycled paper products are also used as fuel in waste-to-energy manufacturing plants. In 1990 there were 135 mass burn plants in operation across the country, "with another 53 under construction or in an advanced stage of planning."[51] However, new plant construction takes time and money. Besides, there is debate about whether or not even the most advanced scrubbers and processes available today adequately filter out the harmful toxins resulting from burning trash for fuel.

The Warmer Bulletin's 1991 *Warmer Factsheet: Fuel From Waste* suggests that incineration be considered only after exhausting source reduction, reuse, and recycling. It recommends assessing both the economic and the environmental aspects of reuse, such as the use of fuel to transport the recovered materials to a processing plant, the energy required, and the pollution caused.[52] We can extend this concept to recycling and incineration since transportation, energy, and pollutants are involved there as well.

In early 1992, leaders of several environmental and community groups co-authored a letter on garbage incinerators and sent it to then New York Senators Baucus and Chafee. In it, they voiced one of their main concerns:

> A 'time out' on the construction of new garbage incinerators is also essential to boost recycling programs. Incinerators burn the same papers, bottles, cans, and other materials which could be recycled. Waste management companies require cities to sign 'put or pay' contracts that require cities to send recyclable materials to incinerators, rather than recycle them. For instance, the city of Babylon, New York, incinerated thousands of pounds of newspapers that were collected for recycling.[53]

> Efficient recycling of high energy-yielding garbage undercuts revenue for incinerators. Indeed, in southeastern Massachusetts, the 1900-ton-a-day incinerator is a big reason why recycling is almost nonexistent in the 32 communities served by Semass."[54]

This controversy may lead to more acceptable alternatives.

City of New York

New York has been promoting recycling for years with its best success in many of the larger businesses. Private haulers had been doing post-pickup MRFing (materials recovery) long before legislation.

Consisting of mostly high-rise buildings filled with office workers, Manhattan generates the largest single volume of waste in New York City. Just how much waste does each worker generate? Although figures from different sources vary, most indicate that the average business generates the equivalent of 0.5 pound of white paper waste per employee per day. In paper intensive companies such as financial firms and educational institutions, that figure can be from 0.75 pound to 1.5 pounds per person per day. And the amount of mixed paper per person is even more. If one person throws out 120 to 360 pounds of white ledger paper per year, then ten people discard enough to fill between five and fifteen postal hampers annually. Extrapolate that to a facility of 500 people, and the figures really start to matter.[55]

Office buildings with employee cafeterias pose another challenge, which is beyond the scope of this book.

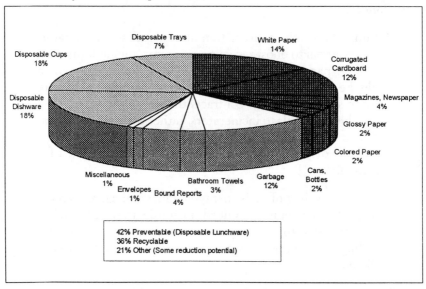

Figure 3-7.   Sample waste stream of office building with employee cafeteria (by volume).[56]

While the city law required recycling, it did not mandate source separation. Consequently, a lot of recyclables were mixed in with real garbage, and lost to the landfills.

The state law was stricter than the city law and required source separation. In the early 1990s, the city brought its requirements in line with the state's regulations and mandated source separation for all businesses in the city, regardless of size or industry. To avoid operational problems, the legislation categorized businesses as either food or non-food establishments. Food establishments must recycle three groups of materials: (1) corrugated; (2) metal, glass, and plastic containers; and (3) flex or film plastics. Non-food establishments must recycle corrugated and paper. Recycling is more complicated for management of multi-tenanted buildings and is beyond the scope of this book.

With mandatory source separation, "commercial businesses must arrange for collection of their recyclables through licensed private carting firms or private recycling companies, not with the NYC Department of Sanitation. Different types of businesses are required to recycle different materials"[57] (see Tables 3-5 and 3-6).

Table 3-5.   Recycling Requirements for Food or Beverage Service Establishments (Restaurants, Delicatessens, Cafeterias, and Bars)[58]

Corrugated cardboard	Flattened boxes
Metal cans	Empty food and soft drink cans
Glass bottles and jugs	Wine, juice, beer bottles; food jars
Plastic bottles and jugs	Soft drink, juice, detergent, milk, water jugs, etc.
Aluminum foil products	Foil wrap, food pans, take-out tins
Bulk metal	Scrap metal: file cabinets, pipes, etc.
Construction/Demolition debris	Including metal, dirt, concrete, rock, etc. Excluding wood, plaster, drywall, wall coverings, roofing shingles, glass window panes

Table 3-6.   Recycling Requirements for All Other Businesses (Office Buildings, Stores, Supermarkets, Warehouses, Manufacturers, Printers, etc.)[59]

Corrugated cardboard	Flattened boxes
High grade office paper	White bond paper: letterhead, typing paper, copier paper, computer printout, etc.
Newspapers, magazines, catalogs, phone books, textiles	If over 10% of the waste stream
Bulk metal	Scrap metal: file cabinets, pipes, etc.
Construction/demolition debris	Including metal, dirt, concrete, rock, etc. Excluding wood, plaster, drywall, wall coverings, roofing shingles, glass window panes

White ledger, newsprint, magazines, phone books, metals, and glass can be commingled, with corrugated set aside for the waste haulers. Trash going to landfills includes most plastics and non-recyclables. The only plastics being recycled are kitchen and cafeteria containers for condiments, salad dressings, and similar supplies. Each company has had to set up internal systems instead of leaving separation to the waste haulers and MRFs.[60]

New legislation went into effect in the summer of 1994, requiring commercial haulers to report the waste they pick up by categories and volume.

Closing the Loop

One common slogan in the recycling arena is: "If you're not recycling, you're throwing it all away." Another slogan that makes a point is: "If you're not buying recycled products, you're not recycling." What good is it if our recyclables just pile up? That's why it's particularly important to buy recycled products made of post-consumer materials. Unless recyclables are used in producing other products, we're just creating a different type of garbage heap.

The supply side of America's recycling revolution has been growing at an explosive rate, but the demand side is still barely under way....In parts of the Northeast and Midwest, where strong after markets haven't developed, the oversupply is so great that state officials report instances of separated trash being mixed together again and hauled to dumps and incinerators at taxpayer expense.[61]

If there is a glut of recyclables, reimbursement rates for generators and haulers will plummet. It is up to us to buy recycled goods whenever they're available and voice our preference for them when they're not.

Sorted mixed office paper typically yields around 90% of the office white, which means more high grade secondary fiber available to be made into a high grade finished product. While many older mills have not, others have retrofitted or started up with newer, more sophisticated technology and more capacity to use recycled fiber. The new technology allows mills to use a wide range of material, especially lower grades. A typical modern mill can process from 500 to 700 tons of office paper per day! This increased capacity to use recycled fiber demands a steady supply,[62] supporting attractive reimbursement rates.

The recycled paper products that modern paper mills produce are often as good as those made of virgin fibers. Those that contain some groundwood may not be the traditional bright white that we're used to seeing, but that doesn't make them any less functional.

To illustrate this point, consider white clothing. People in the United States perceive bright white as a symbol of cleanliness. Yet, the glow of phosphors in the laundry detergent help make white clothing appear white. But in countries where people don't use phosphorous cleaners, white clothing yellows over time, and they perceive white clothing with a yellow tinge as being clean. Similarly, the public's perception of quality must change to accept recycled content as the norm in writing and printing papers.

Source Reduction

Recycling and buying recycled goods can help us make the most of our waste. In order to truly maximize our resources, we must use less of them in the first place.

People growing up in the Great Depression learned to conserve. They saved and reused everything, even gift wrap. They learned to do without and discovered what was really necessary and what was a luxury.

Office paper is the largest component of "waste paper" in the United States today. The New York City case is an example of how costly consumption per employee is to employers. Obviously, waste minimization and source reduction are part of the solution. Are there some simple, inexpensive ways to significantly cut down on internal paper use?

What about electronic mail and the information superhighway? These by themselves may not decrease paper use. In fact, if users generate single-sided printouts of their electronic mail, or documents that would have been printed double-sided for distribution, paper consumption will actually increase. The challenge is getting users to create, distribute, read, and store information electronically and avoid printing whenever possible. When employees effectively use the technology that's available to them, companies can substantially save on paper and disposal costs, directly augmenting the bottom line. What's in it for the employees? They enhance their skills and make themselves more desirable in today's changing and competitive job market.

Electronic mail won't give us the paperless society if we continue to print it. Having that hard copy report instead of reading the same information on-line or printing out that draft instead of editing it on-screen wastes valuable resources.

Bright white desktop paper and hard copies are very familiar to our society. People are always reluctant to give up what is familiar and give up old habits. But if we're to be successful in minimizing waste and protecting our natural resources, then old habits must die. Furthermore, "recycling efforts will not be successful long term if products with a recycled content are not purchased and used by consumers."[63]

But the most compelling reason for businesses to recycle and reduce consumption is purely economic. While companies across the country earn revenue on their recyclables, many report a higher contribution to their bottom lines by reducing waste disposal costs.

In Part II, we'll see how some corporate giants have been successful in changing old habits and have found incremental profits through their recycling efforts. A four-phase step-by-step approach in Part III presents guidelines for an effective corporate action plan that can be adapted to fit the needs of small organizations. Finally, we'll learn in Part IV how we can go beyond paper recycling to reduce our negative impact on the environment.

ENDNOTES

1. *Warmer Bulletin, Number 45* (Tonbridge, Kent, UK: The World Resource Foundation, May 1995), p. 10.
2. *World Resources 1994-95* (New York, NY: Basic Books), p. 311.
3. *1994 Annual Statistical Summary Recovered Paper Utilization: Eighth Edition* (Washington, DC: American Forest & Paper Association, May 1994), p. 48.
4. *Fast Facts: U.S. Paper Recovery Recycling & Reuse* (Washington, DC: American Forest & Paper Association, 1993).
5. *Recovered Paper Statistical Highlights, 1995 Edition* (Washington, DC: American Forest & Paper Association, July 1995), p. 2.
6. *Educator's Waste Management Resource & Activity Guide* (Sacramento, CA: California Department of Conservation, Division of Recycling, April 1992), p. 83.
7. *50 Simple Things Kids Can Do to Recycle* (Berkeley, CA: EarthWorks Press, Inc. and California Department of Conservation, 1991), p. 15.
8. *Educator's Waste Management Resource & Activity Guide*, p. 84.
9. Depending on transport distances involved in taking collected waste paper to mills for re-processing, the methods used to de-ink the paper, and other factors.
10. *The Warmer Bulletin, Number 43, Information Sheet: Paper Making & Recycling* (November 1994), p. 3.
11. About 35 feet tall.
12. *The Warmer Bulletin, Number 43, Information Sheet: Paper Making & Recycling*, p. 2.
13. Based on a gasoline yield of 60%, or 50.4 gallons, and a car that averages 25 mpg. Other by-products such as diesel and heating fuel, jet fuel, lubricants, and petroleum coke would also be saved by recycling. Extrapolated from statistics provided by 76 Products Company, Brea, CA.
14. *Educator's Waste Management Resource & Activity Guide*, p. 83.
15. Based on 160,000 BTUs per hour; at $0.60 per therm, saves $318.
16. Southern California Air Quality Management District, Diamond Bar, CA.
17. *The World Almanac and Book of Facts 1986, 118th Year Special Edition* (New York, NY, NY: Newspaper Enterprise Association, 1985), p. 635.
18. *50 Simple Things Kids Can Do to Recycle*, p. 22.
19. *The World Almanac and Book of Facts 1986*, p. 650, 669.
20. *Educator's Waste Management Resource & Activity Guide*, p. 83.

21. *Recycling for California State University* (San Marcos, CA: Mashburn Waste & Recycling Services).

22. *Educator's Waste Management Resource & Activity Guide*, p. 83.

23. *A Week With Waste: Activity Packet for Teachers* (Sacramento, CA: California Integrated Waste Management Board, 1992), p. 4.

24. Based on 212,000 cubic feet per second.

25. *Webster's Ninth New Collegiate Dictionary*, "Weights and Measures" (Springfield, MA: Merriam-Webster, Inc., 1985), p. 1338; *The World Almanac and Book of Facts 1986, 118th Year Special Edition*, p. 533.

26. *The Warmer Bulletin, Number 42* (August 1994), p. 4.

27. Estimates range between 30% and 55%.

28. Pacific Gas & Electric, San Francisco.

29. *World Resources 1994–1995*, p. 217.

30. *Solid Waste Digest, Vol. 5, Number 7* (Alexandria, VA: Chartwell Publications, July 1995), p. i.

31. Elizabeth Holtzman, Comptroller, *Burn, Baby, Burn: How to Dispose of Garbage by Polluting Land, Sea and Air at Enormous Cost* (New York, NY: Office of the Comptroller, City of New York, NY, January 1992).

32. *Educator's Waste Management Resource & Activity Guide*, p. 81.

33. *Recycling Means Business in California* (Sacramento, CA: Californians Against Waste Foundation).

34. *Facts at a Glance* (Sacramento, CA: California Integrated Waste Management Board), p. 4.

35. *CAW Progress Report* (Sacramento, CA: Californians Against Waste Foundation, Summer 1993), p. 2.

36. Victor A. Bell, Chief, Office of Environmental Coordination, Rhode Island Dept. of Environmental Management. House Hearing on State and Local Recycling Programs before the Subcommittee on Transportation and Hazardous Materials, Committee on Energy and Commerce, April 24, 1991.

37. *1994 Annual Statistical Summary, Recovered Paper Utilization* (Washington, DC: American Forest & Paper Association, May 1994), p. 48.

38. *U.S. Industrial Outlook 1994, 35th annual edition* (Washington, DC: U.S. Dept. of Commerce, January 1994), p. 10-3.

39. Ibid.

40. Ibid., p. 10-4.

41. Ibid., p. 10-3.

42. *1994 Annual Statistical Summary Recovered Paper Utilization, Eighth Edition* (May 1994), p. 48.

43. Ibid., pp. 93–94.

44. *50% Paper Recovery: A New Goal for a New Century* (Washington, DC: American Forest & Paper Association, 1993).

45. *1994 Annual Statistical Summary: Recovered Paper Utilization, Eighth Edition*, pp. 93–94.

46. *12 Facts About Paper Recycling* (Washington, DC: American Forest & Paper Association, 1989).

47. *The Warmer Bulletin, Information Sheet: Paper Making & Recycling* (November 1994), p. 3.

48. *1994 Annual Statistical Summary, Recovered Paper Utilization,* p. 12.

49. Ibid., p. 81.

50. Ibid.

51. *Warmer Factsheet: Waste Incineration* (Tonbridge, Kent, UK: The World Resource Foundation, January 1990), p. 2.

52. *Warmer Factsheet: Fuel From Waste* (Tonbridge, Kent, UK: The World Resource Foundation, January 1991), p. 1.

53. Letter to Senators Baucus and Chafee, members of the Subcommittee on Environmental Protection to support amendments to Resource Conservation and Recovery Act. Phil Clapp, Legislative Director Clean Water Action; Daniel J. Weiss, Washington Director, Environmental Quality Program, Sierra Club; Virginia De Simone, Advocacy Chair, League of Women Voters of the U.S.; Susan Birmingham, Environmental Advocate, U.S. Public Interest Research Group (USPIRG); Marchant Wentworth, Legislative Representative, Isaak Walton League; Rick Hind, Legislative Director, Greenpeace; Carolyn Hartmann, Staff Attorney, USPIRG.

54. *Garbage Magazine,* March/April 1991.

55. 1 postal hamper = 18 bushels = 1 cubic yard.

56. *Annual Report 1993: The Council on the Environment of New York, NY City* (New York, NY: Mayor's Office, 1993), p. 12.

57. *How to Recycle or Reuse Almost Anything,* (New York, NY: NYC Department of Sanitation, Bureau of Waste Prevention, Reuse and Recycling), p. 1.

58. Ibid.

59. Ibid.

60. Materials recovery facilities.

61. "Piling Up: As Recycling Surges, Market for Materials is Slow to Develop," *The Wall Street Journal,* January 17, 1992.

62. Interview with Dan Sandoval, ed., *Fibre Market News,* Cleveland, OH, March 27, 1995.

63. Thomas M. Henderson, Project Director, Resource Recovery Office, Broward County, Florida, April 24, 1991, House hearing on State and Local Recycling Programs before the Subcommittee on Transportation and Hazardous Materials, Committee on Energy and Commerce.

PART II:

WHO'S GOTTEN OUT OF THIS MESS?

CASE STUDIES

CHAPTER 4

AT&T

Background

Recycling is by no means new to blue chip corporation, AT&T. They started recycling over 50 years ago. Back then, customers leased telephones from AT&T and returned their phones when discontinuing service, replacing a damaged unit, or upgrading to newer equipment. Telephones that could not be refurbished and re-leased were disassembled for their parts made of precious metals, which were then melted down and reused. This telecommunications giant now recycles equipment, scrap metal, glass, aluminum cans, wooden pallets, laser printer cartridges, and, of course, paper.

AT&T's facilities in Basking Ridge and Bedminster, NJ, were the first to recycle computer and white bond paper when they opened in the early 1970s. Burke Stinson, District Manager at Basking Ridge, said that back then the company was redeeming over 20 tons per day nationwide. This resulted in over $1,000 in incremental revenues and savings of $3,000 in garbage fees per day—a net gain of $4,000 per day or over $1 million per year![1]

Major Successes

Over the years, the average AT&T office worker has been consistently generating about 1.5 pounds of paper waste per working day. This includes the high volume generated by those who work in the data and records centers and smaller amounts from operations people. In 1990, AT&T's U.S. locations used a total of 60 million pounds of paper. That's 30,000 tons.

At the annual stockholders' meeting that year, CEO Bob Allen announced the corporation's goal to curb domestic paper consumption 15% by the end of 1994. Did AT&T achieve this goal? In fact, employees exceeded it, reducing consumption 29% to a little more than 23,000 tons (see Figures 4-1 and 4-2).

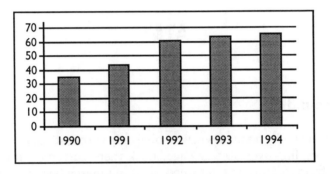

Figure 4-1.   AT&T total paper recycling (% of waste).[2]

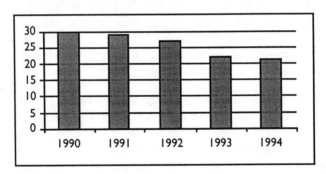

Figure 4-2.   AT&T paper use (in thousands of tons).[3]

Global Real Estate (GRE) historically is responsible for 50% of this paper consumption. Other business units such as Bell Labs and Global Information Systems account for the rest of the paper consumption.[4]

Another corporate goal set in 1990 was to increase the total paper recycling rate from 20% to 35% by 1994. When this was achieved in only one year, the company decided to increase its recycling rate goal to 60% by the end of 1994. Even with this increased target, the company not only achieved but exceeded its paper recycling goal a year ahead of schedule. By the end of 1993, AT&T was recycling 63% of its paper domestically.

In 1991, AT&T's company-wide paper recycling program included white, colored and mixed grade paper, computer paper, envelopes, carbonless forms, magazines, newspapers, and corrugated boxes.[5] However, many sites were recycling only desktop white, computer paper, and some cardboard. In just three short years, the progress was staggering (see Table 4-1).

Table 4-1.   AT&T's Paper Products[6] Recycling Results

	1991[7]	1994[8]
Number of AT&T Facilities Worldwide	2,540	2,500
Number of Employees Worldwide	275,000	304,000[9]
Paper Usage	N/A	29% less than in 1990
Total Volume Recycled (tons)	22,681[10]	24,540
Recycling Rate	45%	65%

To stimulate the market demand for secondary fiber, approximately 98% of the company's direct mail and sales literature had recycled content by the end of 1994. Recycled paper purchases company-wide also increased ten times 1990 quantities. Of course, these impressive numbers wouldn't have been possible without the participation of many dedicated employees. In the pages that follow are the stories of some of the star players.

Bedminster's Dynamic Duo[11]

AT&T's paper recycling program blossomed in its home state, New Jersey, partly because of the state's serious waste disposal dilemma. The company joined a coalition of businesses throughout the state whose mission was to encourage business recycling. After implementing its own successful recycling program, AT&T produced a guidebook and videotape to help carry out this mission and show coalition members and other businesses how they could do the same.

Cheryl LaPerna is AT&T's subject matter expert in recycling and waste management. She first became acquainted with recycling in 1984 while working for Jerry Twardy. Jerry was then (and still is as of this writing) property manager at one of AT&T's largest facilities located in Bedminster, New Jersey. In 1984, at the direction of the company's upper management, Jerry, Cheryl, and people involved in purchasing and facilities management at a few other sites formed the AT&T Recycling Team. Together they set out to expand AT&T's recycling efforts.

All of the company's New Jersey facilities were then recycling only desktop white paper, with receptacles in just the copier rooms. AT&T wanted to expand the program to include mixed paper. However, many local contractors were interested in only computer greenbar and white paper. They were willing to pick up the other grades, too, for a fee.

One of AT&T's records storage buildings was then (1984) utilizing the services of Giordano Paper Recycling Corporation (GPRC) for its document destruction and recycling requirements. Anthony Giordano, Sr., President of GPRC, met with the AT&T Recycling Team and introduced his "If you can tear it, don't trash it" program. He offered to pick up all paper grades including booklets, file folders, hanging folders, window envelopes, desktop white, mixed paper, junk mail, newsprint—everything. Anthony didn't charge for hauling the lower grades of paper and became the first New Jersey contractor to pay AT&T for this type of material.

Because this was unheard of at that time, Cheryl and other Recycling Team members thought that the concept was simply too good to be true. The Team decided to conduct a market search, visit other recyclers, and put out a "request for quotation" for recycling service. In the end, it was apparent that GPRC was the only vendor capable of providing everything that AT&T needed:

- Reliable on-time pick-up service for the 100-plus AT&T locations in the state,

- Document security according to AT&T requirements, and

- Reimbursements at top dollar for mixed office papers (which comprised over 85% of the paper waste generated).

In the ten year period from 1984 to 1994, the Giordano Paper Recycling Company met and exceeded their obligations, paying AT&T $300,000 over and above the required contract negotiated prices for paper and cardboard! GPRC's track record has earned the hauler a top spot in the recycling arena. Because of the relationships that it has developed with its customers since 1984, GPRC has become the exclusive recycler for virtually all of the Fortune 500 companies in New Jersey and Pennsylvania.[12] It has earned the same exclusivity with hundreds of other companies in New Jersey, New York, and Pennsylvania ,thus setting the standards for their competition.

Jerry and Cheryl credit AT&T's recycling success to two key factors: the cooperation of the property managers at each facility (they can make or break any program in their building), and long-term partnerships with vendors like Giordano in New Jersey. "No one stands alone when it comes to recycling," says Cheryl. "AT&T's 'best practices' for recycling acknowledge that throughout the country, our recycling success depends on the ability to partner with first-rate vendors who share our commitment to quality and can meet AT&T's requirements for integrity, reliability of service, and information protection, and pay competitive prices for our recyclables."

As a Recycling Team member, Anthony Giordano, Sr. recommended the two-can system, whereby every employee would have one lined bucket for trash and another unlined one for all waste paper (staples, paper clips, rubber bands, etc.). The loads are picked up, sorted at Giordano's facility, then transported to domestic paper mills where the paper clips and staples are magnetically pulled out and sold as scrap metal. In Anthony's opinion, a white-only paper recycling program would achieve, at best, a 30% recovery rate.

At first, Cheryl thought that two cans per employee weren't necessary, expecting employees would recycle faithfully without a lot of hand-holding. Instead, she placed large bins in central locations throughout the Bedminster facility. Employees were given legal size orange vinyl folders to store waste papers until they could drop them in the central bins during their daily routines. Using this approach, volume increased from 4.3 tons per month for only desktop white to 12.5 tons per month for commingled mixed, white, and computer papers.

The facility also put in a baler to handle the large volume of cardboard. Together, both programs captured enough materials to decrease the amount of pulls of the 40-yard trash compactor. As a result, the Bedminster facility was able to avoid $178,000 in annual waste hauling and disposal costs.

In 1987, two years into the contract with Giordano, Cheryl's department became part of the newly formed Contract Services Organization (CSO). The reorganization resulted in duplication of many job responsibilities. With Jerry's assistance, Cheryl transferred to Records Management in CSO as their first official full-time recycling coordinator. In this position, Cheryl was able to help all of the property managers throughout New Jersey in their recycling efforts. In 1990 her responsibilities expanded to recycling and waste management for AT&T worldwide. She has become the Technical Support Representative for recycling and waste management in what is now called the Global Real Estate (GRE) organization.

By 1989 the central bin program was still in place, but things were about to change. In the habit of walking around the building on informal inspection every day after regular work hours, Jerry would scan the trash cans. He saw the large amount of paper still being thrown away and recommended that Cheryl implement the two-can system. But Cheryl wasn't ready to give up on the central bin plan yet.

So Jerry asked her to follow the garbage truck and take pictures to see how much paper was still being dumped. Cheryl agreed and told him, "I'll bet my paycheck that no more than 10% of that garbage is paper. (But then) I decided that maybe I wouldn't take any pictures, because 50% of the load was still paper."

Seeing all that paper still being trashed convinced Cheryl to introduce the two-can system. Jerry Twardy recalls what happened. "In early December, we gave everybody the option…three ways of doing it. They could have their individualized cans, they could have their folder, or they could have the centralized cans in their area. A lot of them adopted the two-can method. If you have room, the two-can system is probably the more efficient way of doing it."

By end of January 1990, their volume of recycled paper doubled to over 25 tons per month. In some months, it was much as 40 to 50 tons.

The two-can system significantly reduced contamination, too, resulting in more consistent loads and higher reimbursement amounts. The convenience of the two cans encouraged employees to recycle everything.

Company policy stated then, and still does, that paper with highly sensitive information must be shredded and placed in the recycle bin. Shredders used to be located throughout the building. Employees would shred all sorts of printed matter indiscriminately, more than 50 hampers full each day. This cost the company a lot in employee productivity. Jerry changed that. He moved all the shredding machines to a single location where employees could send their confidential papers for shredding. Jerry estimated the volume shrunk to about 12 hampers full per day, less than one fourth of previous amounts. This saved the company a lot of employee time and did not require an increase in staff.

Incentives and Controls

It has been easy to monitor the program. Giordano gives Jerry certified weight tickets showing the number of pick-ups, the content, and the amount of money reimbursed for recyclables. He also provides quarterly reports with detailed breakdowns that show amounts of landfill space, trees, and barrels of oil saved through their recycling efforts (see Table 4-2). AT&T pays for only those materials that it sends to the dump. Since Giordano landed the AT&T account, waste hauls from the Bedminster facility to the dump decreased from three times a week in 1984 to once a week in 1994.

Company executives set goals for all levels of management. Property managers' goals include reducing waste and consumption, increasing the recovery rate of recyclables, and improving customer satisfaction. Degree of achievement is part of the performance review. While Jerry wouldn't do anything differently if this incentive didn't exist, other property managers who may not be as altruistic have cold cash reasons to accomplish environmental goals.

Informing Employees

In orientation, new hires receive a welcome package and employee handbook. Figures 4-3a and 4-3b show the text on recycling and waste minimization procedures from that handbook.

Table 4-2.   Paper Recycling at AT&T–Bedminster, NJ[13]

Year	Total Tons Recycled	Paper Type	Program Type	Disposal Costs Avoided
1984	42.9	White CPO	Central collection bins in copier rooms and some common areas	$ 5,360
1985	104.7	Mixed[14]	Central bins in copier rooms and common areas	$13,100
1986	126.0	Mixed	Orange folder Central bins in copier rooms and common areas	$15,750
1987	128.9	Mixed	Same as in 1986	$16,100
1988	143.9	Mixed Cardboard[15]	Same as in 1986	$18,000
1989	353.1	Mixed	Two-can under-the-desk[16]	$44,150
1990	441.1	Mixed	Two-can under-the-desk	$55,150
1991	438.3	Mixed	Two-can under-the-desk	$54,800
1992	432.7	Mixed	Two-can under-the-desk	$54,100
1993	390.0	Mixed	Two-can under-the-desk[17]	$48,750
1994	437.5	Mixed	Two-can under-the-desk	$54,700

RECYCLING

Contact
    Rich Phelps
    Jerry Twardy

In accordance with AT&T Policy, each employee is responsible for properly disposing of all recyclable materials while on Company premises. AT&T's recycling program in New Jersey includes office papers, corrugated boxes, glass, aluminum cans, scrap metals, and drycell batteries.

At the Bedminster facility employees are provided with the most convenient method for recycling office papers via individual under-the-desk recycle cans. Aluminum cans should be placed into special aluminum can recycling receptacles that are located in the vending and cafeteria areas, as well as in most conference rooms. Any glass bottles or aluminum cans that are left on employees' desks will be picked up by the evening cleaning personnel. Corrugated boxes are also picked up by the evening cleaning crew for recycling.

Papers that are recyclable include: AT&T general proprietary documents, computer and white bond papers, colored papers, cellophane-windowed envelopes, magazines, newspapers, booklets, and carbonless forms.

In order to comply with New Jersey's recycling mandate and Company policy, Building Services cannot empty an individual's recycle or trash can if any materials have been improperly disposed into it.

Figure 4-3a.   Recycling policy from AT&T employee handbook.

WASTE MINIMIZATION

In order to cut down on costs and decrease the amount of waste being disposed into landfills, AT&T has instituted a waste minimization program at all Company locations. The waste minimization program includes the following:

1. Employees are urged to print two-sided copies whenever possible to virtually cut the amount of papers used in half.

2. Jumbo rolls of toilet paper and paper toweling have been installed that use much less packaging materials, and the toweling generates less paper waste in its use.

3. The use of laser printers generates less paper waste in their printing process than the continuous feed printers.

4. Using electronic mail greatly cuts down on paper.

5. The packaging used for AT&T's products has been replaced with smaller, recyclable boxes.

6. Reuse binders, tab, Pendaflex folders, manila file folders, paper clips, rubber bands, and other office supplies as many times as is practical. These items make up a large part of the waste stream and should not be disposed of prematurely.

7. Participate in the Environmental Mug Program (save 500 disposable cups per employee per year, when used!).

8. Reuse interoffice envelopes until all of the address boxes are filled in.

9. Use disposable tableware only for "take out." Use non-disposable when eating in the cafeteria.

10. Buy smart when ordering supplies, bear in mind whether the items can be reused or are easily recyclable once they are no longer usable. Also look for products that are made from recycled materials such as papers and plastics.

• Before purchasing anything, check to see if the items are available used or as surplus stock.

• Binders with the plastic inserts on the cover and spine can be customized and reused easily.

• Use mechanical pencils, and refillable pens.

• Avoid pre-printed binders, brown Kraft and colored papers, carbon paper, or any papers coated with wax or plastic including glossy papers. These items are very difficult, if not impossible to recycle in some areas.

• Remember, white paper is the most easily recyclable.

11. Avoid handouts at meetings. Use vu-graphs or make copies only upon request.

• AT&T Proprietary documents can be recycled without shredding since they are released to an approved recycling vendor. However, papers that are highly sensitive, including those marked AT&T Proprietary (Restricted), and AT&T Proprietary (Registered) must be shredded prior to recycling.

Figure 4-3b. Waste minimization policy from AT&T employee handbook.

Public Relations and GRE communicate results in company news-letters and in an annual environment and safety report. In the Building Management Team (BMT) meetings, representatives from every department find out how they're doing and report back to their colleagues. The volumes of recycled trash are translated into terms that are meaningful to employees, such as money earned, costs decreased, resources saved (trees, oil, energy), pollution reduced, and charities supported.

Obstacles

When it comes to employee buy-in and participation, assembly-line workers in Jerry's facility haven't been a problem. They follow a set process so that implementing recycling, waste reduction, and use of recycled goods in their jobs is easy.

Jerry's challenge is office workers who bring things from home and throw them into the dumpsters near the facility's garage. Their plastic garbage bags have trash and recyclables mixed together. People discard newspapers, magazines, glass jars and bottles, aluminum, various packaging, non-standard materials, even used motor oil. Sometimes his staff will find an envelope or a letter that will help identify the culprit. Then Jerry calls the employee and reminds him/her of the company policy. Fortunately, these contaminants can be easily pulled out of the dumpsters to make the load acceptable. Used tires are his biggest problem because he must pay about $3 each for an authorized recycler to take them. Once he even found someone had discarded the truckbed of an old family pickup.

Obstacles inside the building are handled a little differently.

As mentioned earlier, customer satisfaction is one of the performance review criteria, so everyone in Facilities is careful to be diplomatic in conveying and carrying out company procedures. Before advising someone of their noncompliance, Facilities makes sure that the employee is aware of the policy.

At first, noncompliance was high. By late 1994, only two or three people per day did not comply in a facility of more than 2,500 employees. Janitors follow explicit instructions not to pick up an employee's recyclables or garbage if the wrong items are in the wrong bins. Then, they leave the employee a note explaining the reason (see Figures 4-4 and 4-5). They also give Facilities a list of noncomplying workers the next morning. If one of these employees complains to Jerry, he offers to have that person's garbage separated with a charge-back to his/her department for the labor cost ($20.70 per hour). This gives each department manager responsibility for employee compliance.

Occasionally, Jerry hears, "I don't have the time to recycle." His answer is always, "You don't have the time not to. We won't empty your garbage if you don't separate it." Human nature being what it is, there will always be some people who won't comply and will try to circumvent the system. They may swap their full trash cans for a cooperating employee's emptied one. Since the janitors keep track of whose cans are not emptied because of contaminants, it has been fairly easy to uncover who is costing the company money.

WE ARE SORRY

The janitorial staff was unable to empty your trash can because it contained recyclable materials.

New Jersey's recycling mandate forbids the disposal of materials such as papers, aluminum, and glass into the waste stream.

Materials that may be thrown into "Garbage Only" trash cans include:

Air Freight Envelopes
Carbon Paper (Shred if the imprinted information is of a Proprietary nature)
Cellophane
Damaged Binders (Recycle the papers)
Foam Cups, Plates, and Trays
Food Waste
Metal Waste (Broken Scissors, Staplers, etc.)
Plastic Forks, Knives, and Spoons
Rubber Products
Tobacco Products
Used Paper Cups, Plates, and Towels
Waxed Paper
Writing Implements (Old Pencils, Pens, Markers, etc.)

Contact Building Services regarding on-site recycling procedures for other materials (Aluminum, Cardboard, Glass, etc.)

Thank you in advance for helping AT&T abide by New Jersey's law!

Figure 4-4.   Janitor's note to noncomplying employee about trash.

WE ARE SORRY

The janitorial staff was unable to empty your recycle can because it contains non-recyclable materials.

New Jersey's recycling law requires the separation of garbage from recyclables. Materials that <u>cannot</u> be thrown into recycling cans include:

Air Freight Envelopes
Carbon Paper (Shred if the imprinted information is of a Proprietary nature)
Cellophane
Damaged Binders (Recycle the papers)
Foam Cups, Plates, and Trays
Food Waste
Metal Waste (Broken Scissors, Staplers, etc.)
Plastic Forks, Knives, and Spoons
Rubber Products
Tobacco Products
Used Paper Cups, Plates, and Towels
Waxed Paper
Writing Implements (Old Pencils, Pens, Markers, etc.)

Contact Building Services regarding on-site recycling procedures for other materials (Aluminum, Cardboard, Glass, etc.)

Thank you in advance for helping AT&T abide by New Jersey's law!

Figure 4-5.   Janitor's note to noncomplying employee about recyclables.

An Ounce of Prevention

What do the janitors do if the bin is full before they finish their rounds? Will they throw into the ordinary trash the paper that should have been recycled? Cheryl advises, "Make sure the vendor is reliable so that when you call them for a pickup, they arrive within 24 hours or that could happen." Before the contract with Giordano when they had the white paper program, "we had a couple of smaller vendors, and there were a few times that happened," Cheryl recalls.

The facility was filling 14 to 15 hampers per day and needed pickups every other day. Occasionally, someone would request a large hamper for a day or so when purging files. There were other days when "we oversubscribed to recycling...so we ended up with big mounds of paper and no place to put them," said Jerry. "We also recycle the shredded paper which is in plastic bags (that are) so heavy you can't even pick them up—300 pounds or so." That's after reducing the total tonnage by 75 to 90% of previous levels!

Jerry's building used to have a baler for cardboard. Giordano replaced the cardboard baler with a compactor to store both cardboard and paper outside of the building. Hauls are down to just one per week, even with the occasional "oversubscribing." Jerry's staff manages the volume of waste paper and cardboard more efficiently and frees up the holding areas within the building.

Duplex Copying

Reprographics Services is a service offering of Global Real Estate Property Management. GRE provides office services for itself, Corporate Information Technical Services (CITS), Bell Labs, Global Information Solutions (GIS), and other business units. GRE's current policy for central reproduction services is to fulfill all duplication requests double-sided unless single-sided duplication is specifically requested. Before this became the standard, duplication jobs used to be done as submitted. When the originals were single-sided, so was the finished product. If they were double-sided, the copies were too. Only about 20% of all copies made were double-sided. How did they change that?

Property Management at each site is encouraged to establish lower two-sided copy rates as an incentive for more two-sided copying. In addition to saving the customer money, it also reduces paper consumption. For example, the rate might be $.025 per image (side) for double-sided copying, and $.03 per image (side) for single-sided.[18] Now double-sided copying consistently comprises around 30% to 40% of all duplication jobs.[19]

Some "quick-copy" self-service copiers are equipped with meters that charge the proper department's account when an employee enters his/her department number and authorization code. These machines have higher maintenance costs and require more labor to service and restock than central reproduction machines. The meters help track volume while controlling costs, consumption, waste, and abuse of copier privileges, although personal use of company copiers isn't a big problem. The meters are mainly for billing purposes. Instead of departments being billed based on headcount, they're billed based on actual usage.

Cheryl says, "It's important to have copiers that make double-sided copies easily" to encourage efficient copying and save on paper costs. Of course, simply having the right copiers and a differentiating cost factor won't guarantee goal attainment. People need to promote and support the program, and George Perry is one of those people.

George is Assistant Manager of the GRE Technical Support Organization and an expert on reprographics services. He represents GRE on a team that promotes two-sided copying. George advises property managers on the company policy and helps them adhere to it.

"Paper usage is affected by a variety of factors," says George. First, as the cost of paper goes up, usage will go down because departments watch expenses closely. Second, the campaign for double-sided copying and lower paper output builds awareness, although when campaign momentum slows, copying is not in the forefront of people's minds and consumption starts creeping up again. Also, having the right equipment and making sure the copiers are working properly affects paper usage. Of course, the personal convictions of those who influence purchases impact what copiers the company will buy. Finally, the Reprographics centers providing the services have autonomy in determining the price incentives for double-sided vs. single-sided copying.

George relates how the company tried to make double-sided copying the default standard for self-service copiers. A trial program was conducted in the quick-copy rooms of the Bell Labs facilities. All the copiers were reprogrammed to make double-sided copies the default, but with poor results. If users didn't consciously choose single-sided, they got double-sided, which they often didn't want. Waste paper volume actually increased until finally they went back to the old standard.

Then in 1994, Public Relations funded a campaign for posters, with a seasonal theme, encouraging double-sided copying. With support from George, GRE property managers have cooperated by displaying posters in the quick-copy rooms of buildings with 25 or more copiers. Every three months, Public Relations sends replacement posters with a different theme for the new season. Some of the posters use peel-and-stick cut-outs that workers can rearrange into different scenes, making the posters more personal.

What else has George been doing? He is on a personal quest to make double-sided copying even easier and more pervasive. He has been talking with copier manufacturers about adding another level of sophistication to new copier designs: If the user does not select two-sided and presses the Start button, the copier will not copy, and the display prompts the user to designate his/her choice. So far, the copier companies have received George's suggestion with skepticism, very little enthusiasm, and reasons why it can't be done or won't help curb consumption. George isn't giving up, though, and as technology improves, his vision may become a future standard.

Closing the Loop

In 1991, AT&T printed its annual report and its proxy statement on recycled paper for the first time. These projects took almost 1,000 tons of recycled paper stock, stimulating the market for recycled materials. The company knows that supporting all aspects of a project assures success.

If Cheryl LaPerna would hear anyone taking credit for the company's initial support of "buy recycled," she'd set the record straight. In her files is a copy of a 1986 letter from Jerry to Purchasing urging them to buy more recycled products, especially paper. That's where consumption and waste were highest. Jerry knew that closing the loop by buying recycled goods would support the price structure on the front end. At that time, most of the recycled paper products that the company could use were significantly more expensive than virgin products. It took only a few years for that to change. In the meantime, buyers looked more closely at recycled goods when the prices and quality weren't much of an issue compared to those of traditional products.

George's department has been buying more than half of all the company's copier and printer paper, usually with at least 10% post-consumer and 20% pre-consumer recycled paper (see Table 4-3). Converting everyone to recycled stock was another priority that came from the 1990 stockholders meeting, although it didn't happen as fast as the company wanted or expected.

"People don't want to pay a higher price for an inferior product," said George. He found that the shorter recycled fibers would leave dust in the machines.

George worked on getting his suppliers to lower their production costs. But in the past, recycled fibers cost more than virgin, imposing a

Table 4-3.   Paper Brands on Contract at AT&T in 1995

Manufacturer	Brand Name	Recycled Content
International Paper	Relay	20% post-consumer
Hammermill	Savings DP	15% post-consumer
Georgia Pacific	Geocycle	15% post-consumer
Unisource	Copy Saver	10% post-consumer

5% to 10% premium on recycled stock. Many of his industry sources told him that the market will never achieve parity between recycled and virgin paper prices. His costs for both kinds of paper climbed in 1995 although the gap between them didn't widen. With all paper costing more, George is now concerned that demand for recycled paper by the various departments might drop.

Could George buy recycled paper by the truckload for volume discounts that bring prices close to or the same as that of virgin paper? Yes, but then he'd have to store it, and the cost of storage space might offset any savings.

Not everyone is willing to or can pay the difference. AT&T sees it as part of the price of protecting the environment. The cost of not doing it is so much higher.

As of early 1995, Reprographics Services was also using recycled toner cartridges for laser printers. At half the price of new ones, they helped the company curb expenses, and customers didn't notice a difference in the quality of their printouts.

Beyond Paper Recycling—Paper or Polystyrene?

For a while, the cafeteria at AT&T–Bedminster used the paper hot cups with the fold-out ears. Jerry thought they didn't insulate well. He would see people taking two and three cups stacked to keep from burning themselves, so they went back to polystyrene.

Ordinarily, coffee costs $.65, $.80, and $1.05. The cafeteria gives its customers a $.05 per cup discount if they use their own mugs. If they use disposable cups for coffee, water, or even ice, they pay the extra $.05 for each cup.

Marriott, the food service vendor at AT&T-Bedminster,[20] helped discourage the use of both paper and polystyrene disposables by giving away 16-oz. insulator mugs, which now sell for $1.25. In each cup, there was a note explaining the philosophy behind the program: "Marriott Corporation has designed this RECYCLE CUP to encourage you to join this refill program for beverage purchases in order to cut down on paper usage and waste. Marriott will continue to be an active participant in the long-range planning process at the local and national levels to ensure that our environment and economy are protected for future generations."

Building management tried to recycle polystyrene containers and plastic utensils from the cafeteria, but found it to be a health hazard because of the food contamination. "We even went to the reusable hard plastic cups (tumblers)," says Jerry, which employees would return with plates and trays for washing and continual reuse. "I think that every employee now has a complete set of plastic tumblers at home. We tried the stainless steel flatware, and every employee must also have a matched set of those, too, because we go through about 240 pieces per (calendar) quarter....People (must) either take them home, throw them out, or put them in their desks." Still, AT&T stands by its convictions and replaces missing utensils.

Cheryl prefers to avoid disposable food service containers and utensils. However, when they must be purchased and disposed of under controlled conditions—to take back to a worker's desk or to eat outside on a nice day—she recommends polystyrene. "Controlled conditions" means that they'll be thrown away in central bins and sent to a landfill.

The cafeteria still uses polystyrene. "You can never get away from disposables 100%, and where they are used, polystyrene is a great product," says Cheryl. "In most circumstances where a building is hours away from a polystyrene recycling facility, it is more environmentally friendly not to recycle it. After all, consider the fuel and pollutants involved in transporting a truckload of polystyrene hours away. Compare that to the fact that in modern incinerators, poly burns clean and hot, so it helps to burn other materials that don't burn so readily. And if landfilled, poly is an inert substance. That is, it does not break down and cause the leachates and methane problem that many biodegradable materials cause in the landfill." Cheryl bases her information about polystyrene on data from Walter Boyhan, a highly regarded environmental engineer at AT&T. "There have been studies published by the polystyrene manufacturers and rebuttals from opposing industries, including the paper cup industry," says Cheryl. "There are some inaccuracies apparent in all of their studies. But until and unless other information becomes available, polystyrene surely seems to be the most environmentally friendly disposable on the market today, that is, when you have to use disposables."

She feels differently about commercial take-out foods. In those situations, she'd rather see paper and aluminum containers used because many people are careless or uncaring. They litter the roads, lands, and waterways with non-biodegradable food containers. Polystyrene breaks into smaller pieces that fish, birds, and other wildlife mistake for food.

"One Person's Trash is Another Person's Treasure"

Forming long-term strategic relationships with best-in-class vendors sometimes allows AT&T to negotiate additional services beyond contract requirements. For example, Anthony Giordano subcontracted another company to pick up commingled glass, cans, and plastic bottles from over 40 AT&T buildings at no additional cost to AT&T. Giordano foots the bill for this service, in support of the relationship. Employees can discard glass, cans, and plastic bottles in the same receptacle.

Some contaminants do get in, but workers at the transfer station sort them out of the recyclables. "Even though you'll never get 100% as long as you have human beings in a building," Cheryl says, "chances are that if you have three central containers placed side by side—one for commingled recyclables, one for trash, and one for paper—you'll get substantial compliance. The central cans system works best when it supplements the individual under-the-desk system for collecting papers and trash."

AT&T locations with adequate space at the loading dock area also recycle food waste. They have arrangements with local pig farmers who are properly licensed to accept certain kitchen food waste such as salad preparation food waste, egg shells, and coffee grounds.

When someone vacates an office, very few items end up in a landfill as long as there's someone available to sort it. Office supplies go back to Office Services for re-issue. Giordano subcontracted with other suppliers to pick up scrap metal, wood, and plastics from AT&T and other companies. There's even a furniture reuse program.

When the common areas in the Bedminster building were recarpeted in the late 1980s, the company was able to salvage over 5,000 square feet of carpet tiles in reusable condition. Employees had the opportunity of buying them at fifty cents per tile, roughly $2 per yard, netting the company over $10,000 (and the employees got a bargain). This program is still in place although on a smaller scale.

Aluminum Cans for Burned Children

Some employees put aluminum cans into a separate bin to donate the proceeds to a local charity. They collect approximately 4,400 aluminum cans per week, earning over $5,500 per year. The money goes to the Aluminum Cans for Burned Children program (ACBC) at St. Barnabus Hospital, Livingston, NJ, for non-medical supplies.

Kill-A-Watt Campaign

GRE's efforts go well beyond recycling. Reducing consumption also means reducing energy use and costs. In 1982 when Jerry first started at AT&T, his facility was equipped with 90-watt floodlights in every hallway. The walls were painted charcoal gray. Just a few simple, easy changes did a lot to reduce energy costs. Jerry had the walls painted off-white to reflect more light than the old gray walls. The old ballasts have been replaced with a new energy efficient type and use 5 watt "wattenizer" bulbs. Half of the lights are turned off at 6:00 P.M., the rest at 9:00 P.M. In the morning, half of the lights go on at 7:00 A.M., and the other half at 8:00 A.M. Air conditioning is on from 7:00 A.M. to 6:00 P.M. If an employee wants any of these turned on outside of standard operating hours, his/her department is charged for the wattage.

The inside temperature is maintained at 72–78° in the summer and 68–72° in the winter. The Bedminster building uses a heat plate exchanger instead of running a refrigeration unit. Jerry says, "It's like using free air."

Conference rooms and private offices have light sensors installed. Some light switchplates have stickers on them that say, "Turn me off when I'm not needed. Save energy," or "Turn off when not needed. Conserve energy please."

The janitors used to turn off computers that employees left on all night and over weekends and holidays, but they got complaints. Some users, for example, were waiting for data transfers. Now as a courteous, gentle reminder, they leave a business-card size note that says, "Energy is being wasted here. Kill-a-Watt. Please turn off lights and equipment when not in use."

Personal Commitment

Jerry's concern about the environment extends beyond the work setting. An avid fisherman, he takes his used plastic fishing line to tackle shops and sporting goods stores that recycle it. "What goes up must come down," and that goes for balloons. Jerry has scooped them out of the water on several fishing excursions because "marine life like turtles will eat them and die." He uses all the family's food waste to compost his garden. He even picks up cans littering the street and puts them in his recycle bin at home. Whether at home or at work, Jerry believes in protecting Earth's future. "Lost revenue is secondary to the environment."

"Obstacle" Is Not in Her Vocabulary

Cheryl has had no real obstacles, with strong commitment by property managers who make the program easy for employees to carry out. "In some states, it might be because it's the law, but it's mostly because of the commitment from upper management."

Does Cheryl spread the gospel outside of work, too? "Like most of us, I do take my work home with me," she says. "I'll admit that my family and friends sometimes have to get me off the subjects of recycling and waste management. But more times than not, they are the ones who can't wait to tell me about their wonderful 'at-home' recycling programs. In New Jersey, community recycling efforts are aggressive and include a wide variety of materials. Consequently, everyone I know is recycling a large percent of their waste and bragging about the small amount of trash that actually winds up sitting on their curbs each week."

Local Leadership

In 1995, AT&T Bell Laboratories received an award from the New Jersey Department of Environmental Protection for "Outstanding

Achievement in Recycling." During the past five years, seven of Anthony Giordano's customers have been recognized this way. (Giordano himself was presented with the Leadership in Industry award in 1988.) Cheryl, Jerry, George, and people like them can help change our consumptive society to one that values its resources and uses them wisely. They are true leaders for a better tomorrow.

AT&T Power Systems, Mesquite, TX

Background

About forty miles east of Dallas is the city of Mesquite and the headquarters of AT&T's Power Systems Division. This international microelectronics division has branches in England, Taiwan, China, and Mexico. It makes power supplies for all sizes and brands of computers around the world.

Since 1985, recycling and waste reduction have broadened significantly from their humble beginnings. Everyone in the plant can take credit for the milestones and performance, with special recognition to a few individuals for some key programs.

Shredded Proprietary Paper

Marilyn May joined the Power Systems Division in 1972. During 1989, a suggestion program prompted investigation into the possibility of starting a formal paper recycling process. Through her investigation, Marilyn found that the plant was paying a contractor to shred its proprietary paper in a large shredder, which jammed often because of the humidity and took time to unclog. The shredded paper was then stored in an open-top dumpster, at the mercy of the wind and weather until the waste hauler's weekly pick-up.

Marilyn also found out that corporate security had approved a Dallas recycler who would pay the company to pick up its mixed waste paper and recycle it. Balcones Recycling, Inc., formerly All Waste Paper Recycling, Inc., was bonded and insured to keep the confidential material secure. Balcones could tell AT&T where the paper was at all times until it was shredded and delivered to the paper mills. This eliminated the cost of the shredding contractor and earned the company money on its recycled paper. The contract with the recycler had not yet been signed, so Marilyn went into action. She helped close the deal and began working closely with the vendor and her coworkers.

Cardboard

AT&T was ahead of its time in the early 1980s with its attempts to recycle cardboard. The facility had a compactor from these early endeavors, which were abandoned because revenue was too low for the effort.

In 1991, cardboard accounted for about 65% of the plant's trash. So Marilyn contacted a cardboard recycler who reimbursed the facility $25 per ton, but charged $125 per 5-ton pick-up. "Some months we received revenue for the cardboard, but most months hauling fees far exceeded the revenue."

Marilyn recalls, "During 1993, the city of Mesquite stated that if we gave them (our) cardboard, they would haul (it away) at no charge to us and the revenues would go to the city coffers. AT&T saw this as an opportunity to work with the community and agreed to the proposal."

Faced with the choice of free hauling and no reimbursement or reimbursement that would create a net loss, the facility gave its cardboard to the City of Mesquite. This program ended in 1994 when reimbursement rates more than doubled and AT&T began earning money for its cardboard. While the program was in place, however, the city benefited from AT&T's extra tonnage, and the company avoided incurring extra costs and contributing to the landfill dilemma.

A Team Sport

Also, in 1991, upper management in manufacturing asked the team to write a quality improvement story. They wanted the story to explain the reason for and main emphasis of the recycling activities, which were to decrease the amount of solid waste going to the landfill. At that time, they were sending thirty-seven 42-cubic yard compacted containers and twelve 30-cubic yard open-top containers to the landfill each month. That was a total of 1,914 cubic yards per month, or 22,968 cubic yards per year, enough to fill 6,960 family size pickup trucks!

Another initiative came from Receiving. A lot of the solid waste was wood from new and used shipping pallets. With help from the team, Richard Vickers, a worker on the receiving dock, volunteered to separate new and old pallets. He also had the recycler provide a full-time trailer at the dock, which he and his colleagues filled with broken pallets, thus earning money for the company. Because they were still throwing away quite a bit of wood, Richard found other vendors to take the scrap pallets as well as crates and wood spools.

Manufacturing personnel also got involved. Many of their supplies came on spools made of cardboard and plastic and in plastic tubes packed in polystyrene foam peanuts or foam blocks. When the spools were empty, workers cared enough to actually separate the cardboard from the plastic and put them into separate bins for recycling. This added to their cardboard volume and helped start their plastics recycling efforts. It was discontinued because the work involved in disassembling the spools created a safety issue.

The team reported the following initial successes (see Table 4-4).

Table 4-4.   AT&T-Dallas, TX: 1991 Recycling Volume

Recycled Material	Pounds
Cardboard (Jan.–Dec. 1991)[21]	260,000
Polystyrene foam blocks (Oct.–Dec. 1991)[22]	492
Polystyrene foam peanuts (Oct.–Dec. 1991)	115

With everyone working together, from 1991 to 1994 they reduced the volume going to the landfill by 20,232 cubic yards and extended the lives of the materials they diverted (see Table 4-5 and Figure 4-6).

Table 4-5.   Cubic Yards of Trash Sent to the Landfill

	1991	1994
42 Cubic Yard Closed Containers	37	4
30 Cubic Yard Open Containers	12	2
Cubic Yards per Month	1,914	228
Cubic Yards per Year	22,968	2,736
Cubic Yards Rescued from Landfill in 1994[23]	N/A	20,232

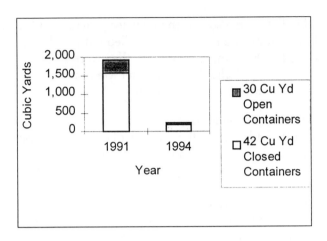

Figure 4-6.   Cubic yards of trash sent to the landfill.

A large portion of this was paper and cardboard (See Table 4-6).

Table 4-6.   Paper and Cardboard Rescued from Landfill in 1994

	1991			1994			Rescued from Landfill in 1994		
	Pounds	Tons	Cu Yds	Pounds	Tons	Cu Yds	Pounds	Tons	Cu Yds
Cardboard	72,286	36.1	115.7	540,994	270.5	865.6	468,708	234.4	750.0
Paper	221,842	110.9	354.9	229,296	114.6	366.9	7,454	3.7	11.9
Total	294,128	147.1	470.6	770,290	385.1	1,232.5	476,162	238.1	761.9

In terms of environmental advantages, the savings are even more dramatic (see Table 4-7).

The price on the open market increased in September 1994, to $50 per ton for cardboard and $55 per ton for mixed paper. Marilyn seized the opportunity and arranged with a paper recycler for pick-ups with no hauling fee. At the full $50 per ton, the company earned $1,400 the next month for their paper and cardboard. By 1995, AT&T Power Systems was mixing all papers in secure containers that the recycler hauled, shredded, and reimbursed. Revenues soared, but not because the volume increased. Rather, reimbursement rates had skyrocketed, yielding much higher returns for the same volume as before (see Table 4-8).

Table 4-7.   Environmental Savings from Recycling 239 Tons of Paper and Cardboard[24]

Environmental savings in 1994 vs. 1991 ...	enough to ...
4,063 trees @ 17 trees per ton	forest 13.5 acres[25]
478 barrels of oil @ 2 barrels per ton of paper	drive an average American car round trip L.A. to N.Y. 54 times[26]
1,673,000 gallons of water @ 7,000 gallons per ton	to flood 5.1 acres with water one foot deep[27]
979,900 kwh of electricity @ 4,100 kwh per ton	supply heat and air conditioning to 121 households for an entire year[28]
12,667 million BTUs @ 53 million BTUs per ton	run a gas clothes dryer non-stop for nine years[29]
765 cu. yds. of landfill space @ 3.2 cu. yds. per ton	fill 239 family size pick-up trucks!
14,340 lbs. of air pollution @ 60 lbs. per ton	drive round trip between New York and Los Angeles 833 times[30]

Table 4-8.   Comparison of Reimbursement Rates in the Dallas Area 1990 vs. 1995[31]

Reimbursement per Ton	1990	1995
Cardboard (65% of trash)	$20	$140
Mixed Paper	$10	$215

By eliminating hauling costs of up to $190 per pull and selling its paper and cardboard, Power Systems Division can now earn a net revenue of as much as $6,237 per month, or $74,844 per year (see Table 4-9). This incremental revenue goes straight to the bottom line.

Table 4-9.   Paper Recycling Revenues 1990–1995[32]

	1990	1991	1992	1993	1994	Jan–June 1995
Cardboard (65% of trash)	$0	$0	$0	$0	$2,426	$13,683
Mixed Paper	$3,040	$1,448	$985	$1,208	$6,721	$12,775
Total Amount Earned	$3,040	$1,448	$985	$1,208	$9,147	$26,458

When office and shopfloor workers dispose of cardboard boxes, they break them down for proper handling by the janitorial service. "Workers are pretty dependable," says Marilyn, and that saves labor costs and increases reimbursement revenues.

An Inside Job

Working with vendors was the easy part. The bigger challenge was selling co-workers on the idea and getting people to change their habits. That's where the Planet Protectors made a difference.

In 1990, the Environmental, Health and Safety Department had an idea for a quality improvement program that included recycling and used a team approach. This gave birth to the Planet Protectors, a team of thirteen members which later grew to fourteen. Marilyn volunteered to be a team member. Within a few months, she became team leader. The Planet Protectors researched the best ways to implement and maintain a recycling program and generate incremental revenue for the company. They looked at what other companies in the area were doing and decided to take it in stages. They focused first on paper, cardboard, and wood, then on aluminum cans and polystyrene, and finally on plastics, spools, and tubes.

They wanted to make the greatest impact as quickly and easily as possible. Therefore, the Planet Protectors went with an all-mixed paper recycling program which included greenbar and high grade white paper.

In March 1991, the team formally kicked off the program by giving each employee a recycling can clearly marked "Recycled Paper Only." At the Planet Protectors' request, the janitorial service hired an extra janitor exclusively to pick up the paper and cardboard. A portion of the hard surface floor in the manufacturing area was set aside for the recycle containers and taped off for easy distinction.

The Planet Protectors wanted to involve everyone—manufacturing staff and office workers—and asked each department and section to designate one person as its recycling coordinator. Marilyn advises that coordinators be people who care and want to take on the extra responsibilities because they believe in recycling, not because it is mandated. Marilyn had experience with another quality team that rotated members every three months. She found that frequent rotation of members prevented smooth transitions. Therefore, the Planet Protectors agreed that recycling coordinators would serve on the team for as long as they chose. When one of them leaves the team, which occurs infrequently, there is usually a willing replacement from the department.

The fourteen Planet Protectors meet weekly to report on progress, follow up on assignments, and receive new ones. They have monthly luncheon meetings with all the recycling coordinators to educate them on recycling and other environmental issues and to keep them motivated. Each coordinator is assigned specific tasks, like encouraging the recycling of laser printer toner cartridges. At the monthly luncheons, everyone proudly wears their "AT&T Recycles" T-shirts and Planet Protector buttons. Marilyn holds a second meeting later in the day for coordinators on the second shift so that no one is left out. Ten additional employees and guests are invited to each meeting. Of course, monthly door prizes help boost attendance. There's always a good turnout— usually 90 of the 130 coordinators. Those who aren't there are on vacation, out sick, or have other commitments.

According to Marilyn, "We really have become a family because of the things that we do. I try to make it real home-like, really interesting."

Marilyn works hard to do just that by bringing in outside experts and community representatives as guest speakers. They cover various subjects such as recycling, hazardous waste, toxic waste, and other environmental topics that are of interest to the workers. The Planet Protectors even show half-hour videos[33] at monthly meetings.

Employees wanted to be at the forefront of the company's environmental efforts. "It's exciting," says Marilyn. "As people found out what was going on, they wanted to help. The program has really grown."

The publicity and progress reports downplayed the revenue earned. Employees seemed to respond better to cost savings, resources conserved, and the impact of their actions on the environment.

"People are doing things on their own. They're bringing us ideas for recycling. They're taking ideas they've learned at work home with them.[34] "Because many employees live in communities that have either limited recycling programs or none at all, they started bringing things from home into the building for recycling—cardboard, junk mail, and telephone books. To make it even easier, AT&T has put bins in the parking lot for cardboard and newspaper. "It's amazing what people will carry from home all the way to this building for recycling."

Videos, charts, and other easily identifiable images like the ones in Table 4-7 help employees understand the volume of waste they've saved from the landfill. "I try to make everyone feel like they're really accomplishing something...for themselves, for their children, for their homes, for their neighbors." And they have!

There is, of course, a cost to keeping a recycling program healthy. When the cafeteria enacted its waste minimization measures, there were one-time expenses of $344 for an aluminum can crusher and storage container and $13,305 for reusable plastic glasses and a dishwasher rack. In 1994, Power Systems spent approximately $8,500 for employee training and awareness, Earth Day activities, the school program, extra containers, and upgrades to existing containers.

Reduce

Duplex Copies

Power Systems believes in source reduction. Every copier in the building can print double-sided copies. Management requires all the TQM[35] materials be printed double-sided. "You're saving on the cost of the actual paper even though you're being charged for each side printed," Marilyn explains.

The Paperless Office

Electronic mail was implemented to reduce the amount of paper generated for internal use. PC users are now in the habit of reading their messages and important documents on the computer screen without printing hard copies.

Phil Lombard has noticed the difference. Up to 1993, he was on a paper use reduction team established by the Data Center. Phil tracked paper usage, consumption, and savings. As the campaign progressed, he saw lower consumption and greater savings.

The Data Center used to use reams of paper printing entire reports. Data Center personnel documented the waste by photographing trash cans full of printed reports. Some had never been used and had the original rubber bands still around them. In 1986, the staff presented management with the Report Management Documentation System (RMDS). The strategy was to have users look at their reports on-line using RMDS and avoid unnecessary printouts.

With the switch to RMDS, those who needed reports could view the data on their monitors. Even when they had to have hard copies, they printed only the pages they needed and only on the days they needed them. Education and awareness efforts paid off.

The 3.7 million images (sides) per month in 1986 would have made a stack as high as the New York World Trade Towers (1,575 feet). If they had been laid edge to edge, they would have covered the distance from Dallas to El Paso (624 miles).[36] The Data Center set a 1991 goal to reduce the 3.7 million images (sides) made per month by 50%. It achieved that goal on schedule, reducing the number of images made per month to 1.9 million in 1991 (see Figures 4-7 and 4-8).

If stacked, the pile in 1991 would have been half as tall (790 feet) as in 1986. If laid edge to edge, the number of copies made would have stretched only as far as Big Springs (320 miles) (see Figure 4-8).

Figure 4-7.   Source reduction equivalent in height, 1986–1991.

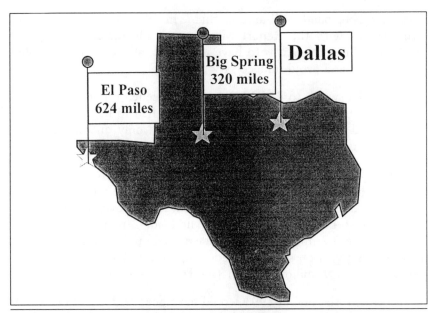

Figure 4-8.   Source reduction equivalent in miles, 1986-1991.

At the end of 1991 the team renewed efforts, setting a new goal to reduce paper consumption another 50% by 1995. Despite a jump in the employee population in 1992, paper consumption decreased fairly steadily. As Figure 4-9 shows, the Data Center accomplished its 1995 paper reduction goal of 50%. The total net decrease from 1986 to 1995 was 73%.

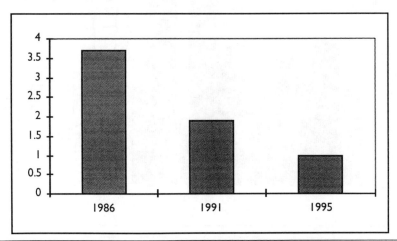

Figure 4-9.   Data Center's paper reduction goal achievement.

Savings extended beyond the cost of paper. The Data Center was able to eliminate one 1200-line-per-minute printer, saving rental charges and reducing operating costs. By going to RMDS, it was able to eliminate hundreds of pages per day and go to less expensive printers. It replaced two high capacity Xerox 9700 laser printers with two smaller ones: a Xerox 4090 printer for the big reports, and a Xerox 4050 for small reports and to serve as back up for the 4090. Although slower, the new printers have been able to adequately handle the reduced volume. The average printing volume per month is 73% lower than in 1986 (see Figures 4-10, 4-11, and 4-12). Fewer hard copies mean decreased costs through lower printer rental fees, which are levied on a per-image basis.

Ron McCauley, team leader for paper reduction in the Data Center, attributes this dramatic drop in expenses to efforts by individual employees to reduce consumption.

But people sometimes must have their hard copies to file. When the Data Center relaxes its campaign, the page count rises. While the big push is on with lots of reminders and focus, the page count goes down

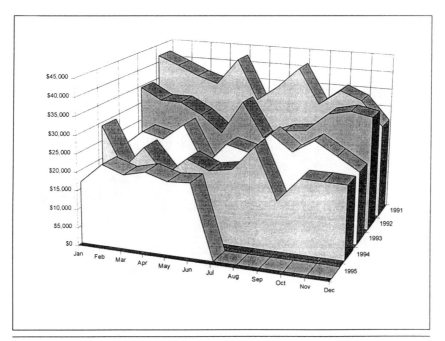

Figure 4-10. Comparison of printing costs by month 1991–1995.

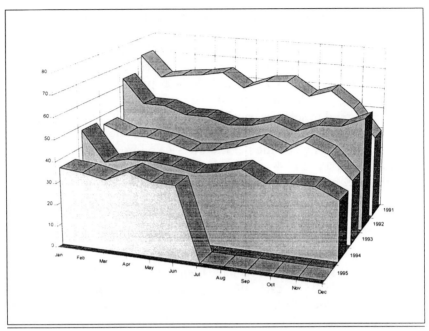

Figure 4-11.   Daily average number of images printed from 1991 to mid-1995.

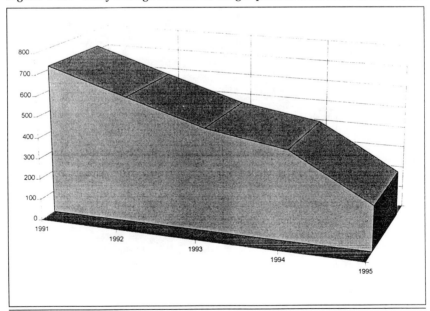

Figure 4-12.   Average number of pages printed per year from 1991 to mid-1995.

with strong response and high compliance. Ron estimates that office workers could reduce the print volume another 25% by putting more types of information on-line.

Overcoming Obstacles

At first, there was quite a bit of hostility because people felt they were being told what to do. A lot of people laughed at Marilyn, saying that her intentions were good but that recycling was a stupid idea.

Others were two-faced. They'd tell her that they thought the program was great. Then they'd turn around and throw their paper in the trash or throw contaminants in with the paper.

The aluminum can recycling bins have holes in the covers clearly marked "Cans Only." Occasionally, Marilyn has seen workers scraping their food trays into them, perhaps spitefully, contaminating the valuable contents.

The department recycling coordinators clean out the contaminants. They then remind errant co-workers of what should and should not go into the bins and reiterate everyone's responsibilities. They help fellow employees realize how worthwhile the program is and invite them to attend one of the training meetings. As coordinators have stood by their convictions, kept after co-workers, and earned respect, the hostilities have subsided.

There is no formal orientation for new hires. With regular meetings, though, and the diligence of coordinators and co-workers, new hires are quickly enrolled in the team effort.

The Planet Protectors were organized as a quality team and assigned the recycling project as a team. "We do a lot with just a little bit of space," says Marilyn. "The time was right for recycling and waste reduction, and management support was very strong." But once the program was past the difficult start-up phase and momentum subsided, the Planet Protectors found it even harder to keep things going. Management considered it in maintenance mode and assumed that it required fewer human resources.

There used to be a whole department for waste disposal. Now that most of the waste is recyclables, there are fewer people involved in waste disposal on even a part-time basis.

Marilyn emphasizes that someone must be put in charge of the recycling efforts full time for the program to succeed. "There's actually more work in recycling than just pitching trash in a trash can and dumping it," Marilyn points out. "Most everything in this facility is recyclable, but you have to find ways to collect and recycle it. You have to segregate it and find people (haulers and recyclers) who are going to take it to designated holding areas. So it's a lot more complicated. Your trash pickups now become recyclables pickups. Companies don't realize that it takes a lot of people to do that. If the company doesn't feel it's important enough to give you the time to do it, you can't do a good job." It takes a lot of time, endless patience, and a great deal of hands-on activities, like loading trucks and pulling contaminants out of the bins.

Without approval authority or direct reports, she has had to depend on a lot of people to get things done. Fortunately, they like and respect her, so she and the Planet Protectors have a very high success rate in everything they do.

"You can't give up." When she'd hear of a department where people weren't following the program, she'd work on the agenda for the next department meeting. She'd talk for a few minutes about their recycling accomplishments, thank them for their patience and cooperation, and show visuals or charts. "You have to get to the people." Some people badgered Marilyn, believing that she was making money for the company at their expense. "Employees don't always realize that a healthier bottom line may save their jobs." Over time, however, they did realize that these efforts meant a better future for their children. They need constant reminders. "You can't let it go. It'll stop if you let it go." Even when she was given additional job responsibilities, she persevered because she is committed and really cares.

In Marilyn's experience, people really want to recycle and reduce waste, but they don't always know how. That's where education and an easy process make a difference.

Marilyn sums up in a single word how she overcame the many obstacles—determination. "There's a lot of frustration because people just don't want to change," she says. "They have to be continually reminded of the rules, not because they don't want to cooperate, but because they forget or other priorities divert their attention."

To others embarking on a similar journey, she offers these recommendations:

- Take the time to research and document everything very well.

- Get the right approvals on all proposals and projects.

- Keep all the figures together.

- Maintain on-going relationships with suppliers.

Closing the Loop

Marilyn and the Planet Protectors have spent a lot of time educating employees and promoting not only recycling, but also closing the loop by buying products in recycled packaging.

Most of the paper products used for copiers and laser printers at Power Systems' Mesquite headquarters are from recycled fiber such as Springhill[37] Incentive DP and Springhill Unity DP. Incentive DP contains a minimum of 50% recycled fiber, with 40% of the total being pre-consumer and 10% post-consumer. Unity DP is 30% pre-consumer and 20% post-consumer. Even the sticky notes from 3M are 100% recycled.

AT&T occasionally asks vendors to change their practices, such as using less tape to make recovery and recycling easier. But the company cannot dictate to their thousands of suppliers that packaging be made from recycled materials. It can, however, control what it ships out and has switched to all recycled packaging, which bear the recycled and recyclable emblems.

Because there's no storage room for office supplies, the division's expense store processes orders from department secretaries on a daily basis. Miller Business Products, its former supplier with locations throughout the United States, carries a full line of recycled office products. By working with Miller Business Products, Power Systems recovered virtually all of its spent laser printer toner cartridges. The expense store took back depleted laser toner cartridges from department secretaries. The cartridges were then returned to Miller Business Products to recycle. However, in 1995 AT&T switched to another national supplier. This new supplier did not have any system in place to take back spent supplies like packaging and laser cartridges, putting the Planet Protectors back to "square one" in some aspects of the program.

Power Systems closes the loop on packaging. Departments return cardboard and shrink wrap packaging from previous deliveries for reuse or recycling. Vendors also support the effort to close the loop by participating in the plants' Earth Day celebration with displays and free samples of recycled products.

An increasing demand for recycled paper and cardboard may sustain higher reimbursement rates. For example, according to Marilyn, Champion Recycling Corporation in Houston, TX, accepts newsprint and magazines, using sophisticated de-inking and recycling technology. It sells the recycled products, primarily newsprint, to newspaper companies throughout Texas. Because Champion is a ready customer for waste paper and sells large volumes of recycled stock, it has helped to stimulate the market in the central southern United States.

The Fort Howard Paper Company in Tuscaloosa, OK, produces Mardi Gras brand paper products, which is 100% recycled. It also markets the Envision line of business paper products—paper towels, toilet tissue, and napkins for the cafeteria. At Marilyn's suggestion, the janitorial service started buying Envision recycled paper for "use-and-disposal," such as toilet tissue and hand towels. In this case, buying recycled was actually cheaper than buying virgin fiber. The company saved over $4,000 in the first year.

Employees use Eberhard's EcoWriter pencils made of recycled materials. The shaft is made of 100% recycled cardboard and newspaper fiber. The eraser barrel is made from recycled aluminum. The EcoWriter is distinguished by a green eraser.

Secretaries can order whatever brands of products they want and buy a lot of items made from recycled materials. The Planet Protectors and their suppliers educate employees on the environmentally friendly products that are available to them (see Table 4-10), and it seems to make a difference.

Table 4-10.   Environmentally Friendly Products at AT&T Power Systems

Item	Recycling	Buying Recycled
Plastic bubble pack	100%	100%
Paper	100% all	high grade white
Cardboard	100%	100% cardboard
Laser cartridges	100%	none
Toilet tissue	none (of course)	100%
Paper towels	none (of course)	100%
IC tubes	100%	none available
Miller's Office Suppliers' delivery boxes (until 1995)	100% return/reuse	100% return/reuse

Reuse

On the shipping and receiving docks, workers flatten containers and return them to suppliers for use in subsequent shipments.

Recycling coordinators don't emphasize removing and reusing rubber bands, paper clips, butterfly clips, or manila and hanging folders. But Marilyn has noticed workers doing this automatically, perhaps because of an increased awareness of wasteful habits that are changing over time.

Everyone's a Winner

During the month, the 14-member team audits the trash for contaminants and gives out grades for performance. Each month, the department or section with the best recycling performance is awarded a roving trophy that stays in the department all month. The department name is added to a special plaque that is in the display case.

Near the main cafeteria is a display case with awards attributing successes to departments. Marilyn points out that these awards are given to all employees as a team, not to individuals or to AT&T. The company is a recognized member of the National Paper Recycling Project. The Planet Protectors design bulletin boards with environmental messages that change every month to maintain momentum.

In 1993, Power Systems received former governor Ann Richards' Environmental Excellence Award for their recycling efforts and reduction of hazardous waste. The plant also won the Environmental Vision

Award from the Recycling Council of Dallas. It's not Marilyn or the Planet Protectors who won but all the employees. Thus, recognition and all awards are presented to the employees.

The Powerful Ideas program awards points to employees whose cost saving ideas are implemented. Employees accumulate points and choose gifts from a catalog. In fact, employees have individual "powerful ideas" quotas that are part of their performance evaluations. Hourly employees who may not have the expertise to come up with technical ideas contribute ideas on how to protect the environment. While it takes formal experience to come up with technical solutions, anyone can contribute ideas on how to protect the environment.

In addition, each month the Dallas team recognizes the employees of the department contributing the most to the plant's recycling efforts by awarding them its Recycling Program Trophy. The division has won many other distinctions over the years (see Figure 4-13).

Clean Texas 2000 Partner

Distinguished Proud Partner (Mesquite)

Dallas Corporate Recycling Council Member

National Office Paper Recycling Project

1992 Power Systems Monthly Recycling Award

1993 Power Systems Monthly Recycling Award

Environmental Achievement Award

Celebration of Excellence President's Award

Recycle America plaque

Special Christmas Donors

AT&T Champion of the Environment (for Community Service)

Figure 4-13. Power Systems Division environmental awards won.

Giving Back to the Community

Power Systems tried to follow the model set by AT&T facilities in Atlanta. There, AT&T trucks go to the company's various locations and collect waste paper for recycling. The trucks stop at several other organizations along the truck routes and pick up their waste paper to help these community members with their recycling efforts. The company calls this "hubbing."

Autistic Treatment Center

The Atlanta hub concept (see details in the section on Atlanta) was replicated in the Dallas area but fell apart when the bottom dropped out of the recycled paper market in the early 1990s. Instead, the Dallas offices teamed up with an autistic children's group, The Autistic Treatment Center. Jim Van Orden, AT&T Director, Public Relations, initiated a plan for AT&T to donate funds for the Center to purchase a truck. With properly licensed drivers, the children go all over Dallas, picking up waste paper and other recyclables from the smaller AT&T facilities. Autistic people are well suited to repetitive tasks such as sorting high grade and computer greenbar paper from mixed paper. By doing so, they maximize their earnings. They make money, take pride in their special skills, and have a responsible place in the society. At the same time, they eliminate storage problems for local AT&T offices while AT&T provides jobs, income, and support for the community.[38]

Toys for Tots

Aluminum cans were already being collected in cardboard boxes before the formation of the Planet Protectors. Allen Cooperider, who joined the team in 1990, had boxes throughout the building. He posted signs informing workers that he would buy toys for tots with the proceeds.

Simply putting out clearly marked cardboard boxes got people to pitch in. All the employees worked very hard to keep the plan clean. When the Planet Protectors team was established, team members wanted to reward their co-workers' efforts with attractive collection containers. They replaced Allen's cardboard collection boxes with sturdier containers lined with plastic bags. The team also arranged for Allen's supervisor to let him collect the cans from the 60 or more containers every other night during his evening shift.

Employees know that every penny goes into a special savings account and that at the end of the year, it all goes to the company's Toys for Tots program. All fourteen team members personally shop for the best buys on toys, bring them into the plant, and show all the employees what their recycled cans have bought. Then they deliver the toys to several carefully researched nonprofit organizations in the greater Dallas area. In 1990, the cans they collected brought in about $600. In 1994, the special savings account totaled $3,000 by donation time.

Now, there are even aluminum can recycling bins outside the building for employees to bring their cans from home. Every Monday the containers are full with cans people have brought from home. As with the cans collected inside the building, the redemption money goes to the company's Toys for Tots program (see Table 4-11).

Table 4-11.   Money Redeemed from Aluminum Cans for Donations

Year	Money Redeemed
1990	$600
1991	$1,330
1992	$1,620
1993	$1,187
1994	$2,864
5-Year Total	$6,001
Annual Average	$1,500

Working for Peanuts

The plant gives smaller companies other items for reuse such as used polystyrene peanuts. This helps these companies financially and extends the life of the items.

Back to School

At one time, the company-approved loose leaf binder was a four-ring type. Now everyone uses standard three-ring binders which are more popular and therefore more easily reused. What have they done with the obsolete four-ring binders? They donated them to the local schools. "We donate as much as we can (that isn't proprietary) to the schools."

Thinking Globally, Acting Locally

Employees have become more active on a local level, too. Many participate in neighborhood cleanup programs and Texas Recycles Day.

Learning from Each Other

Businesses in the area have formed the Dallas Corporate Recycling Council to promote recycling and exchanging information. As an AT&T representative, Marilyn is on the Council's Board of Directors. She learns from others in the local business community, shares her knowledge and experience, and helps implement resource-saving and cost-cutting ideas.

The Mesquite facility shares ideas with other community members, too. It "adopted" the elementary school across the street. The children participate in AT&T's Earth Day activities and special projects. They also bring their recyclables to the plant, dump boxes of waste paper into AT&T's bins, and receive badges that say, "Rugel Elementary School and AT&T Recycle Together."

Marilyn periodically gives talks and seminars to the children on how AT&T and the community can protect the environment. She uses visual examples to help them understand and think about how much damage people have caused the world. For example, "If every person in the world were allotted a patch of ground 2 feet-by-2 feet on which to stand, the world's population would fit into an 800-square-mile area about the size of Jacksonville, FL." That's a fairly small area, and yet we have done enormous damage to the global environment.

Visual images like this get the children to think about the effect—both negative and positive—that they can have on their surroundings.

In 1993, Marilyn initiated an annual, company-sponsored program for the eighth graders. It was a science project named "Solutions for the Earth." Planet Protector team members visited the school, graded the science projects, gave cash awards to winners, and presented gifts to all the students who participated. They brought the students back to the plant and gave them a tour. "Touring our high-tech factory was a real treat," says Marilyn. "It gave these rural students the opportunity to see what big business is like and to expand their career ideas." After the tour, they celebrated—with cake, of course.

Beyond Paper and Cardboard

Marilyn frequently receives questions on environmental issues and recycling other materials such as vehicle oil and plastic bottles. As a result, she's always learning new ways to reduce waste from vendors, research, and fellow employees. Everyone shares what they've learned through meetings, newsletters, and Earth Day celebrations.

Earth Day

Power Systems celebrates Earth Day with a different theme every year. In the early years, the emphasis was on areas such as preserving the rainforests and curbing ozone depletion. Another year it was on endangered animals. Still another year, the focus was on household hazardous waste.

Hazardous Waste

AT&T already has programs in place to recycle nickel cadmium batteries (found in many portable computers). When employees asked how they could recycle their spent alkaline batteries, the battery recycler arranged to take them as well. Employees now bring their dead household batteries to their department secretaries who then give them to the company's battery vendor.

The Gift of Sight

Even eyeglasses are recycled. The plant has its own eye care center in the building. Employees can turn in their old eyeglasses for redeployment by the Lion's Club. Lion's Club volunteers arrange for the glasses to go to underprivileged countries where doctors give them to patients who would otherwise be unable to afford them.

Everybody into the Pool!

Carpooling was scheduled for 1995. With 2,400 employees in the building, Marilyn planned to have a social gathering and get them to meet each other. Like a lot of people, "they won't just pick up the phone and call someone they don't know. We will do a match-up of home addresses first. Getting acquainted with someone first makes it a lot easier to find someone who's compatible." Van pooling was also on the agenda of new programs.

Plastics

Shop floor workers used to put plastic packaging in large hard-shell containers. These were difficult to pick up and consolidate into the main bin. Then, one of the shop workers, Keith Vincent, suggested a free-standing open-frame container using a durable bag that could be conveniently pulled out when full, easily emptied, and reused.

To reduce and eventually eliminate polystyrene cups, each employee receives a reusable cup that says, "Power Systems Cares—We Recycle". Since April 1994, polystyrene cups have been completely eliminated from the entire plant, saving $15,000 in the first year.

The cafeteria started recycling its cardboard and plastic containers in 1992. When added to the rest of the plant's recycling, this has brought waste disposal costs way down (see Table 4-12) and saved landfill space.

Table 4-12.   Waste Disposal Savings 1991–1994 (in cubic yards, except where noted)

	1991	1992	1993	1994	4 Year Total	Avg/yr	Avg/yr/ employee	# pick-up trucks full/ employee/yr
Sent to Landfill								
Plant & Factory	7,554	6,546	6,588	5,844	26,532	6,633	3	0.86
Cafeteria	3,936	3,432	1,920	1,920	11,208	2,802	1	0.36
Total	11,490	9,978	8,508	7,764	37,740	9,435	4	1.23
Landfill Savings								
Plant & Factory	15,414	16,422	16,380	17,124	65,340	16,335	7	2.13
Cafeteria	0	504	2,016	2,016	4,536	1,134	0	0.15
Total	15,414	16,926	18,396	19,140	69,876	17,469	7	2.27
Cost Avoidance								
Plant & Factory	$35,760	$45,426	$45,615	$89,664	$216,465	$54,116	$23	N/A
Cafeteria	$0	$900	$3,600	$3,900	$8,400	$2,100	$1	N/A
Total	$35,760	$46,326	$49,215	$93,564	$224,865	$56,216	$23	N/A

The Next Hurdle

As of 1995, the City of Mesquite doesn't take glass for recycling. Power Systems has to send it to the landfill or, to recycle it, would have to pay someone to pick it up. Eventually, the Planet Protectors will find a way to recycle glass, too. Fortunately, the plant has very little glass to discard. Beverages are generally in plastic bottles and aluminum cans.

As of 1995, magazines, telephone books, newspapers—anything that tears—are in the mixed paper recycling program.

Summary

Power Systems has been on a relentless mission to carry out the company's environmental and waste reduction goals. "We do everything here: reduce, reuse, and recycle," says Marilyn. Officially, there are fourteen people on the Planet Protectors team. In reality, everyone in the Mesquite headquarters is an active member of this highly successful team.

AT&T, Chicago, IL

Chicago is home to one of AT&T's largest employee sites. The downtown facility began its paper recycling efforts in the early 1990s. The internal program was similar to the one in Bedminster, NJ, and also experienced a high degree of success. What's even more important is AT&T's leadership role in the community. The company serves as a shining example of how big business can shape the lives of its neighbors.

Internal Leadership

As in Bedminster, when the program started, everyone received a second under-the-desk basket strictly for waste paper. The janitorial service empties the separated waste every night and periodically reports the revenues and avoided costs to AT&T.

An integral part of garnering early support for the program was building employee awareness. This was done through one-hour training sessions. Every employee was expected to attend the small group meetings. Items pulled from AT&T employees' trash were fastened to a rope that was hung in the conference room from one corner to another. The training was thus named "Trash on a Rope." Trainers used this visual symbol during the training classes, showing employees all the things that they were to separate, and what was recyclable and what was not. Employees had fun while they learned, were motivated, and have formed good habits.

Community Leadership

The city of Chicago has ordinances pertaining to waste disposal and recycling. While most of the large businesses already have recycling programs, the smaller ones are often unaware of the ordinance or don't know how to comply. Instead of "hubbing" as it did in Atlanta, in Chicago AT&T has supported the nearby communities by giving financial grants to local environmental organizations.

The program "AT&T Community Recycling: Do It Today for Greener Tomorrows" was initially designed to address the entire recycling loop, from community collection to job creation. In the process, it supported two nonprofit organizations in Chicago and nearby communities and provided support to one of them in attempting to establish a re-manufacturing facility in a Chicago West Side neighborhood. In partnership with these recycling organizations, AT&T has helped educate neighborhood businesses on how to comply with city regulations, follow a viable business model, and learn about and address recycling and disposal issues. Together, they also educated neighborhood residents on waste reduction. As a team, they contributed to community awareness about recycling habits that would help protect the environment.

Recycling in Schools: Resource Center

In 1994, the Chicago public schools wanted to develop recycling programs but did not have the resources, funds, or expertise to administer them. They found that the costs of start-up education, collection containers, and material transport were prohibitive.

That's where AT&T's community partner, the Resource Center, came to the rescue. This organization is "committed to the advancement of environmentally responsible waste management. Its primary emphasis is recycling in urban areas."[39]

With the support of an AT&T grant, the Resource Center has worked with all grade levels of public schools on Chicago's South Side. The Center has helped them build their own recycling efforts, get everyone in each school involved, and make the program self sufficient by collaborating with other schools and businesses. The combined resources of the Center and the schools generate about 10 tons of recycled paper monthly.[40]

Job Creation and Community Development: Bethel New Life

Bethel New Life is a 16-year-old community-based economic development corporation dedicated to creating quality affordable housing and livable-wage jobs, and to providing family and senior services on Chicago's West Side. In the early 1980s, Bethel New Life identified and selected environmental enterprises as growth areas to target for rebuilding the West Side economy. It began a buy-back recycling center that eventually put more than $1 million into the hands of thousands of community residents through the purchase of thousands of tons of recyclable material. Bethel then started a commercial collection program and then developed a material recovery facility (MRF). Bethel transferred the MRF operations to a minority recycling company in 1994 as part of its mission to nurture recycling and minority enterprises.

Bethel New Life teamed up with the Institute for Local Self-Reliance (ILSR) to help provide jobs and expand the tax base in the West Side. ILSR promotes recycling as an essential component of economic growth. Bethel and ILSR work together to attract scrap-based manufacturing companies to the West Side. These companies would consume recycled materials from local recycling centers to produce products. The plan is to develop a recycling manufacturing cluster that creates synergies between waste generating and waste consuming businesses. To date over a dozen companies have expressed interest.

AT&T funded some phases of Bethel New Life's five-phase plan.

Phase 1:
- Develop the basic plan for approaching businesses for neighborhood business expansion.

Phase 2:
- Determine requirements for expansion of its current material recovery facility site to meet the projected flow of additional materials.

- Determine public awareness and outreach to business and government agencies.

- Develop legislative strategies to increase recycling and enhance the relocation of plants to the community.

- Conduct a workshop for community members, local government officials, investors, and entrepreneurs to report findings.

In **Phases 3 and 4**, Bethel New Life created a roster of scrap-based manufacturers, determined their screening and selection process, and prepared a site-specific plan.

Phase 5:
- Prepare a detailed report on the entire project, methodology, and results.

- Circulate the report to economic development and community organizations.

- Conduct a national training conference for bankers, community development agencies, and community corporations for replication in other U.S. cities.

Chicago Recycling Coalition

These groups received AT&T grants in early 1994. In the fall of that year, AT&T additionally awarded a grant to the Chicago Recycling Coalition (CRC). The CRC is a nonprofit organization that promotes source reduction and recycling as one of the most environmentally efficient ways for the City of Chicago to manage its municipal solid waste. In harmony with the work that the two other organizations have accomplished, the CRC informs small businesses in Chicago about the city ordinance on recycling. Through an AT&T grant, the CRC published a free guide to business recycling that was distributed to 6,000 businesses in Chicago. The booklet, along with 40 presentations that CRC made to business organizations, describes cost-effective recycling programs and explains ways businesses can save money using efficient recycling practices.

Summary

As for the success of the internal program, a monthly newsletter informs employees in the downtown location about company events, progress of their own recycling efforts, and updates on the local organizations they support. Annual surveys show that environmental articles are among the most popular.

AT&T's Chicago employees respect the roles they have to play to protect our delicate environment. They are sincerely interested in preserving our precious resources for future generations and do it responsibly and willingly.

More AT&T Successes

Atlanta, GA

AT&T is a community leader in Atlanta. The company's own trucks collect recycled paper from its 47 buildings within a 50-mile radius. It helps other organizations that are along the truck routes and either don't have sufficient volume for local recyclers or would not be able to do their part if not for AT&T. These include several schools in the area, a local city hall, and a police station. The trucks stop and collect their neighbors' waste paper while making their daily rounds. Doing this enhances the economies of scale, qualifying AT&T's loads for the highest possible reimbursement rates.

By "hubbing," the Atlanta operation tripled its volume of recycled paper from 45 to 140 tons per month. The Atlanta Hub recycles paper, cardboard, aluminum, and other materials, diverting approximately 2,200 tons of materials from the waste stream each year. The company makes a net profit of over $500,000 annually from this program. This includes over $300,000 in revenue for recyclables and more than $200,000 in cost avoidance by not hauling the materials away as trash.

Warrenville/Westwood, IL

The Westwood Ecology Quality Circle in the Warrenville/Westwood, IL, facility started its recycling program in April 1991. By the end of the year, it had recycled over 29 tons of paper. It saved 493 trees, 58 barrels of oil, 203,000 gallons of water, 118,900 kilowatts of electricity, and 93 cubic yards of landfill.[41]

Ballwin, MO

The NSEC-West Headquarters, Ballwin, MO, can thank the Environmental and Recycling Quality Team for its leadership in recycling 147 tons of paper. In 1991, it saved 2,500 trees, 294 barrels of oil, 1,029,000 gallons of water, 602,700 kilowatts of electricity, and 470 cubic yards of landfill. Furthermore, NSEC-West Headquarters contributed $8,000 of the money earned from recycling to the community's charitable activities.[42]

Kansas City, MO

The Kansas City Service Center's "Waste Watchers" identified ways to reduce the volume of waste being hauled to landfills. In a multi-phase approach, the team defined and researched the problem, analyzed the research, developed a trial program, conducted a cost-benefit analysis, made the necessary modifications, and implemented the final plan. By working as a team in a methodical manner, the Kansas City Service Center was able to recycle 77% of the 195 tons of waste materials generated annually. It reduced waste disposal costs from $92,000 to $75,000 for 1991. It further offset its disposal costs by earning $7,120 from recycling for a total cost reduction of $24,120.[43]

AT&T EasyLink Services

One way to consume less paper is to employ the information superhighway. AT&T EasyLink Services makes many of its forms, such as vouchers, vendor payment, phone cost reimbursement, purchase request, and several other forms, available in electronic format. Transactions can occur without ever printing a single page.

Total Commitment

AT&T's commitment to protecting the environment goes far beyond paper recycling. It has permeated every pore of the company's business. At the end of 1991, the company already boasted significant achievements because of the company-wide commitment. By the end of 1994, the list of accomplishments was even more impressive (see Table 4-13):

Table 4-13. AT&T Environmental Goals

(excludes Audit & Remediation and Safety)	1991[44]	Base Year	1994[45]	Base Year
Increase Paper Recycling	45%	1990	65%	1990
Reduce Paper Usage	N/A	N/A	29%	1990
Reduce Manufacturing Process Waste	39%	1987	66%	1987
Reduce CFC Emissions	76%	1987	100%	1987
Reduce Total Toxic Air Emissions	73%	1987	96%	1987

To put 1994 in perspective, AT&T:

- Nearly doubled the paper usage reduction goal.

- Exceeded the goal for paper recycling.

- Decreased manufacturing waste disposal by more than double its goal of 25%.

- Eliminated chlorofluorocarbon (CFC) emissions 19 months ahead of schedule.

- Practically eliminated reportable air emissions.

We've already seen some of the hands-on activities that made paper and cardboard recycling a way of life at AT&T. What did it take to realize these other impressive accomplishments?

Reducing Manufacturing Process Waste

AT&T research teams travel to its factories to understand "the environmental impact of manufacturing processes and operations."[46] Team members conduct systems analyses "with waste minimization audits and process modeling to provide a 'cradle-to-grave' view of waste. They track the waste stream from the raw material brought into a factory, to the products that go out the door, even to recovery of those products at the end of their lives."[47] The goal is to reduce waste while increasing product output yields.

CFC Phase-Out

Scientists have found CFCs to be damaging to the ozone layer that protects living organisms on Earth from the sun's damaging and cancer-causing rays. In fact, television weather broadcasts in southern California now include the projected UV levels along with the forecast smog levels.

For many years, industry used CFCs as solvents in the manufacturing process. In February 1992, President Bush announced that the United States would phase out CFC production by the end of 1995. Even before that announcement, AT&T engineers had already changed many different manufacturing processes in factories around the world. In 1986, the company emitted 2,622,049 pounds of CFCs. That was reduced 76% to 627,213 pounds by 1991 and eliminated in early 1993, almost two years before the target date.

Toxic Air Emissions Reduction

Other toxic solvents used in manufacturing include methylene chloride, methyl chloroform, trichloroethylene, tetrachloroethylene, acetone, methyl ethyl ketone, and perchloroethylene. AT&T decreased 1991 emissions by 73% of its 1988 baseline volume. One plant reduced emissions by 95%, and another one had completely eliminated them. Simply by changing two processes, an AT&T plant in Ohio went from being the state's second largest polluter in 1987 to eliminating the offending pollutant. It saved the company $210,000 as well. "The company achieved these decreased emissions by reducing the use of materials, not by disposing of pollutants through some other medium."[48]

AT&T's goal for 1995 was to reduce reportable air emissions 95% by the end of the year. Because of employees' efforts and innovations, factory emissions were 96% lower than in 1987 a full year ahead of schedule.

The Honor System

AT&T incorporates audit and remediation activities in its performance reporting. The number of audits has doubled, and factory audit frequency is up 60% since 1991. More than 650 environmental and safety audits supplement factory audits. To penetrate the organization, business units and divisions conduct self-audits. Central compliance monitoring, control mechanisms, and Quality Policy Deployment methodology assure AT&T of its leadership position.

This is evident by the long list of partnerships in which it has taken an active role and the many awards it has won. AT&T recently received the American Forest & Paper Association's 1994 Best Paper Recycling Award (honorable mention) and the 1994 Washington Business Recycling award for outstanding achievement.

What can the company possibly do to top such outstanding performance in less than a decade? Set new goals (see Figure 4-14), of course:

A Vision for the Future
AT&T Environmental, Health and Safety Goals for the Year 2000

We will sustain the successes and significant gains we've already made against our goals in CFC elimination, manufacturing waste disposal reduction and reportable air emissions reduction, as we add new operations and as we grow globally...

In addition, by the year 2000:

We will put in place internationally recognized E&S management systems for at least 95% of our products, services, operations and facilities.

We will ensure that at least 95% of our services, operations and facilities meet the rigorous criteria of AT&T's Model Safety Program.

We will develop and apply Design for Environment (DFE) criteria that provide competitive, environmentally preferable products and services.

We will improve the energy efficiency of our operation, avoiding what would otherwise be the emission of at least 500,000 metric tons of greenhouse gases.

We will recycle at least 70% of our wastepaper.

We will continue to use quality policy deployment and methodologies to achieve our goals, and we will engage our employees in addressing environmental and safety issues in the workplace, recognizing their achievements at work and in their community environmental efforts through such programs as AT&T Champions of the Environment, and telecommuting.

Figure 4-14.   AT&T environmental, health and safety goals.[49]

Executive Perspective: Take It From the Top

Changing habits starts at the top. Bob Allen, CEO, practices what he preaches and believes employees should, too. They all have a copy of the company's environmental vision and policy, signed by the Chief Executive Officer himself. In just a few words, they embody AT&T's foundation for the twenty-first century (see Figure 4-15).

William Ebben, Vice President of GRE, is very committed to recycling. In 1987 before recycling was popular, he budgeted and paid for a dedicated person to handle recycling and waste management full time.

On a broader scale, Brad Allenby, research vice president in AT&T's Technology and Environment group, is a proponent of DFE (Design For Environment). He co-authored a book on industrial ecology with Tom Graedel of the technical staff. Allenby and Graedel speak about DFE

AT&T's Environmental Vision	AT&T's Environmental Policy
AT&T's vision is to be recognized by customers, employees, shareholders and communities worldwide as a responsible company which fully integrates life cycle environmental consequences into each of our business decisions and activities. Designing For Environment is a key in distinguishing our processes, products and services.	AT&T is committed to protection of human health and the environment in all operations, services and products. AT&T will integrate life cycle environmental quality into design, development, manufacturing, marketing and sales activities worldwide. Implementation of this policy is a primary management objective and the responsibility of every AT&T employee.

Figure 4-15.   AT&T's environmental vision and policy.[50]

tools and methodologies to companies and environmental organizations all over the world. "We did the easy stuff by making the proper environmental adjustments within the existing model," says Allenby. "Now it's time for the hard stuff—heavy duty, fundamental changes in the company—and we're going to have to break that old model and evolve to a new one in the process."[51]

Final Notes

Sharing the Secrets to Success

Naturally, a company the size of AT&T has a lot of clout and the advantage of volume. But size can also be a detriment. Even the best animal trainer doesn't get a grown elephant to turn on a dime. Yet this mammoth corporate leader has made itself nimble by following a few basic rules which it readily shares with other companies and organizations.

Make It Easy. AT&T's property managers make sure that there are large, clearly labeled paper recycling containers in copier rooms and in high traffic office areas. Many employees receive separate under-the-desk containers for recyclable and for non-recyclable waste. Cafeterias, vending areas, and conference rooms have clearly marked containers for aluminum cans and glass bottles.

Make It Personal. To kick off a new recycling program, employees receive information on what materials are recyclable and the benefits of recycling. Janitorial crews in some facilities leave reminders for noncomplying employees, and may not empty their receptacles.

Use a Team Approach. Three strategies have paved the way for everyone in the organization to get on the band wagon:

- Have role models at the top.

- Sell and market the concepts and programs to all employees.

- Recognize and award achievement.

Follow a Socially Responsible Philosophy. AT&T's philosophy about protecting the environment is rather simple: "It's much more effective—and cheaper—to prevent pollution than to correct it...by designing it out of company products and manufacturing processes.... Process changes often require lengthy—and expensive—approval processes, by the company and by customers.... Our goal is a clean and healthy planet, and we can't relax our efforts."[52]

AT&T realizes that if we don't monitor ourselves, government will do it for us. "Our experience in the United States tells us that it is both environmentally and economically smart to stay ahead of governmental regulation. Even though we complied with environmental requirements in the 1950s and 1960s, some of the approved landfills we used then now require remediation efforts."[53]

"Focusing on design, manufacturing and use from a product's inception to rebirth allows designers to anticipate negative impacts to the ecosystem and engineer them out of the process. This always makes environmental sense, and most of the time makes financial sense as well."[54]

Lead from the Top Down. What does leadership mean at AT&T? It means securing management's commitment to make environmental protection policies part of the corporate goals. It means following a total quality management methodology wherever it does business. It means integrating a Design For Environment approach to manufacturing for "cradle to reincarnation" of every product it creates. It means instituting effective new employee training and grassroots participation. While the

Environment and Safety Engineering team provides engineering support to enact the company's goals, the local facilities implement strategies and tactics to achieve those goals.

Commitment from upper management, a system that's easy to implement and follow, and employees who believe in what they're doing are the keys to this success. AT&T leadership knows what doing the right thing means, and that there are some things that don't have a price tag.

ENDNOTES

1. Based on 260 workdays per year.
2. *An Investment in Our Future: AT&T Environment & Safety Report 1994* (Basking Ridge, NJ: Studio W, Inc., 1995), p. 7.
3. Ibid.
4. Phone interview with George Perry, Assistant Manager, GRE Technical Support Organization, Basking Ridge, NJ, March 1995.
5. *An Investment in Our Future: AT&T Environment & Safety Report 1991* (Basking Ridge, NJ: Laurence Mach Creative Services, Inc. 1992), p. 6.
6. Includes all grades of paper and cardboard.
7. *An Investment in Our Future: AT&T Environment & Safety Report 1991*, p. 6.
8. *An Investment in Our Future: AT&T Environment & Safety Report 1994*, p. 7.
9. As of April 1, 1995; per AT&T, June 1995.
10. 45,362,000 pounds.
11. Personal interviews with Cheryl LaPerna, Assistant Product Manager, Recycle & Waste Management, Contract Services Organization, and Jerry Twardy, Property Manager, Bedminster, NJ, October, 1994.
12. According to Giordano Paper Recycling Company.
13. The population at this location has fluctuated between 2,400 and 3,000 employees throughout the ten year period.
14. As of August 1985.
15. As of October 1988.
16. As of January 1989.
17. Lower tonnage due to reduced consumption efforts such as duplex copying and electronic mail.
18. Average price. Actual price is based on each facility's operating costs, and therefore varies slightly. However, the relationship of costs and savings are similar.
19. Phone interview with George Perry, Assistant Manager, GRE Technical Support Organization, Basking Ridge, NJ, March 1995.
20. Service America is another very cooperative food service company operating in other AT&T locations.
21. Based on an estimated 5,500 to 6,500 pounds per week.
22. While polystyrene does not weigh a lot, it takes up a lot of space. Per-container hauling fees can really add up.

23. Compared with 1991.

24. Reflects increase in tonnage recycled in 1994 vs. 1991; figures are rounded off.

25. Based on 300 trees per acre.

26. Assuming an average 25 miles per gallon and a gasoline yield of 60%.

27. Based on 325,857 gallons per acre foot. *WaterLines*, Vol. 95, No. 9 (Laguna Niguel, CA: Moulton Niguel Water District, 1995), p. 4.

28. The average U.S. residential customer uses approximately 8,100 kilowatt hours to heat and air condition his/her home annually according to Pacific Gas & Electric, San Francisco, and *Educator's Resource and Waste Management Guide*.

29. Refer to Figure 3-1.

30. Based on 2,786 miles. *The World Almanac and Book of Facts 1986, 118th Year Special Edition* (New York, NY: Newspaper Enterprise Association, Inc.), p. 150.

31. Higher rates in 1995 due to increased use of recycled goods by consumers generating higher demand.

32. "1990–1993 cardboard hauling fees exceeded the revenue received. We gave the City of Mesquite the cardboard as a community project. We changed recyclers in 1994 when the market value increased." Marilyn May, July 1995.

33. *Race to Save the Planet Earth* (South Burlington, VT: Annenberg/CPB Collection).

34. *An Investment in Our Future: AT&T Environment & Safety Report 1991*, p. 6.

35. Total Quality Management.

36. Visual equivalents are from AT&T Power Systems and are estimates only to personalize the impact for the reader.

37. Springhill is a division of International Paper Corp.

38. Interview with Jim McMahon, AT&T Public Relations, August, 1994.

39. *AT&T Community Recycling Chicago Market: Concept Paper* (Chicago, IL: AT&T).

40. Interviews with Jim McMahon, Director Public Relations, Basking Ridge, NJ, and Alison Pikus, District Manager, Chicago, IL.

41. *An Investment in Our Future: AT&T Environment & Safety Report 1991.*, p. 6.

42. Ibid.

43. Ibid., p. 11.

44. Ibid., p. 1.

45. *An Investment in Our Future: AT&T Environment & Safety Report 1994*, pp. 4-7.

46. *An Investment in Our Future: AT&T Environment & Safety Report 1991*, p. 12.

47. Ibid.

48. Ibid., p. 15.

49. *An Investment in Our Future: AT&T Environment & Safety Report 1994*, p. 16.

50. Ibid., p. 15.

51. Ibid., p. 12.

52. *An Investment in Our Future: AT&T Environment & Safety Report 1991*, p. 12, 14, 15.

53. *An Investment in Our Future: AT&T Environment & Safety Report 1994*, p. 3.

54. Ibid., p. 2.

CHAPTER 5

McDonald's Corporation

Environmental leadership is part of the basic corporate philosophy at McDonald's Corporation and has been for many years. In the late 1980s, many people in top management across the country might have avoided any involvement with environmental groups, but not Ed Rensi. Amidst controversy about polystyrene food containers, the President and CEO of McDonald's USA started a bold initiative that led to many company and community programs. This chapter describes some of them.

Before going into the details, let's look at McDonald's waste stream to understand the scope of the challenge.

Waste Stream Challenges

Franchisees own and operate over 80% of the more than 10,000 McDonald's restaurants in the United States. A network of close to 600 suppliers provide them with food, packaging, and equipment. In the high volume, low profit-margin fast food business, changes in business practices have to be economically sound as well as socially responsible.

McDonald's Corporation kicked off its Waste Reduction Action Plan in April 1991. While the programs at the other companies featured in this book focused on their office settings, McDonald's waste management challenge revolves around its restaurants.

"We looked at every aspect of McDonald's operations, including materials discarded behind-the-counter and by customers in the restaurant lobby, disposable packaging used in McDonald's take-out business, and the distribution and supply system.... Almost 80% of McDonald's on-premise waste, by weight, is generated 'behind the counter' in the preparation area and restaurant supply system. Many of these waste materials, such as corrugated boxes, are candidates for source reduction and can also be identified and separated for recycling by the restaurant crew."[1] In some cases, reusable alternatives can replace disposables.

A special 1991 Environmental Defense Fund study estimated that, depending on the restaurant, drive-through and take-out customers made up 50% to 70% of McDonald's business, and over-the-counter solid waste accounted for about 21% of the average restaurant's total solid waste. Controlling over-the-counter waste requires a different approach. The company decided it needed to modify its business practices and change designs to reduce the amount of waste that ends up in customers' hands.

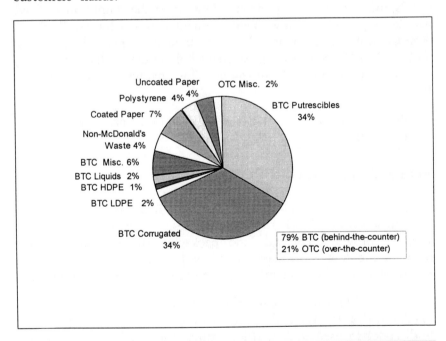

Figure 5-1. **Waste stream of a typical McDonald's restaurant.**[2] (Total Solid Waste = 238 lbs. per day or 0.12 lbs. per customer served).

Meeting the Challenge

At the suggestion of the Environmental Defense Fund (EDF), Rensi met Fred Krupp, EDF Executive Director, to build on McDonald's strong environmental foundation. Together they formulated an environmental corporate mission for McDonald's to reduce its impact on the environment and widen employee and customer awareness..

To accomplish this mission, a task force of four McDonald's and three EDF representatives did extensive research into the company's practices. With EDF's impressive backgrounds in biochemistry, chemical engineering, economics, hazardous waste management, and environmental science, combined with McDonald's operations and packaging experts, the task force members developed a far-reaching action plan. They took care in making sure that their solutions would not have other negative environmental effects.

Shelby Yastrow, Sr. Vice President, championed the company's relationship with the EDF. Shelby was confident that McDonald's and the EDF could find common ground, and they did. The initial Waste Reduction Action Plan encompassed 42 and eventually 100 initiatives, pilot projects, and tests to reduce waste. Bob Langert, Director of Environmental Affairs, worked with Keith Magnuson, then Director of Operations, to implement the program that earned the President's Environmental and Conservation Award. In fact, McDonald's has earned many awards for its environmental leadership (see Table 5-1). Its partnership with the EDF has been a springboard for other environmental partnerships that have had widespread benefits (see Table 5-2).

This worldwide fast food giant has three reasons for recycling:

- Customers expect it.

- The best way to avoid legislation is to make it unnecessary.

- It's good business.

Let's look at each of these.

Table 5-1.   U.S. Environmental Leadership Awards[3]

Fred Schmitt Corporate Leadership Award	1991	Waste Reduction Plan developed with the Environmental Defense Fund
President's Environmental and Conservation Challenge Award	1991	Partnership with the Environmental Defense Fund
Rated #1 by consumers (Cambridge Research International)	1991-1993	Environmental performance
#1 Corporate Environmental Reputation by consumers (Roper Green Gauge Survey)	1991-1993	Environmental reputation
Ameristar Environmental Award (Institute of Packaging Professionals)	1993	Big Mac package (40% recycled content, 10% lighter)
Corporate Social Responsibility Award (Society for the Advancement of Management)	1993	McRecycle USA program

Customers Expect It

With more than 16,000 restaurants in 83 countries, every day McDonald's serves over 28 million customers worldwide. Its research shows that its customers really do care about the environment. It may not be the first thing on their minds, but it is certainly an important one. As the company's Director of Environmental Affairs, Bob Langert, says, "We are a customer-driven business... (Our customers) expect us to take a lead on the environment. Just as they expect clean bathrooms, they expect us to be environmentally responsible."[4]

Before implementing the Waste Reduction Program, surveys showed that 53% of McDonald's customers felt good about its packaging. With the program in action and the switch from the polystyrene clamshell to paper based wraps, that changed to 88% by the Spring of 1991.[5]

Thinking Globally, Acting Locally

"Starting in 1955, Ray Kroc (McDonald's founder) picked up litter for several blocks surrounding his first restaurant."[6] The restaurants carry on the tradition. "From the start McDonald's demonstrated environmental leadership by directing crew members to pick up litter within a one block radius of the restaurant.[7]...Being a business leader carries the responsibility of being an environmental leader as well."[8]

Table 5-2.   McDonald's Environmental Partnerships[9]

Smithsonian Institute	1989	Rain Forest Trayliners to increase consumer awareness
World Wildlife Fund	1989	5 million WEcology magazines for youths
Keep America Beautiful	1989-today	Leadership member
Environmental Defense Fund	1990-91	Waste Reduction Action Plan
Global ReLeaf	1991	Over 9 million trees given to customers
Conservation International	1991	"Discover the Rain Forest" Happy Meal booklets to educate young patrons
Buy Recycled Business Alliance (National Recycling Coalition)	1992-today	Founding member; dedicated to increasing the purchases of recycled products
Student Conservation Association	1993-today	Created McDonald's All Star Green Teens high school environmental recognition and educational program
U.S. EPA's Green Lights Partner	1993-today	Lighting energy conservation program
Environmental Defense Fund	1993-today	Member of Paper Task Force to buy more environmentally preferable paper products
Conservation International / Clemson University	1993-today	Restoring rain forest land in Central America's La Amistad Biosphere Reserve ($4 million project involving several public and private sector organizations)
U.S. EPA's WasteWi$e	1994-today	Voluntary waste reduction program
National Audubon Society	1994	April 1994 "Earth Days" Happy Meal
National Audubon Society / The Composting Council	1994	Participant in foodservice industry's composting initiative "Food for the Earth"

Community Education

The company's Communications Department has produced several environmental videos for both adults and teenagers. Restaurant owners/ operators can use them in local outreach programs to educate their communities. Other educational materials range from explanatory in-store brochures for customers to workbooks and teacher's guides for use in schools.

Calling All Phone Books

In February 1992 McDonald's published its "Blue Book" on implementing a local/regional phone book recycling program. The restaurants made arrangements with local telephone companies and waste haulers to pick them up for recycling. Later that year the San Diego region kicked off its Calling All Phone Books Program in conjunction with 19 stores in the eastern and southern portions of San Diego County. McDonald's provided a free 16-ounce soft drink to every person depositing a Pacific Bell SMART Yellow Pages directory at participating restaurants. More than 730 tons of directories were collected for recycling and kept out of the landfills.

> "Recycling has become one of McDonald's cornerstone business philosophies, embraced by every McDonald's restaurant throughout the county," said Michael Dill, senior operations consultant at McDonald's Corporation in San Diego. "In addition to our support of recycling programs such as 'Calling All Phone Books,' we provide an end market for recycled products such as napkins and bags."[10]

Shaping Tomorrow

In 1994, McDonald's ran the McWorld Stamp Design Contest in all its U.S. restaurants. Targeting children between the ages of eight and thirteen, it was an opportunity to increase environmental awareness in tomorrow's leaders. The McWorld Stamp Design Contest was a joint effort with the U.S. Postal Service to design four U.S. postal stamps for 1995 release. Four winners had their designs become regular 32¢ postage stamps, an event marked by a ceremony in Washington, DC, on the 25th anniversary of Earth Day.

The Best Way to Avoid Legislation is to Make It Unnecessary

The company has chosen to stay ahead of the legislative curve and seek solutions before encountering problems.

McDonald's went beyond the task force's initiatives. It formed an Optimal Packaging Team with representatives from all pertinent departments. The Team's charter was to continually evaluate and explore packaging alternatives to improve performance, as well as environmental and other traditional aspects.

To share ideas with its owner/operators, McDonald's publishes a pocket guide, *Fifty Ways Your McDonald's Can Help The Environment.* It lists 24 tips on conserving energy, 11 ideas for conserving natural resources and educating consumers, 10 ways to be a "waste buster," and 5 suggestions on remodeling for the future. Many of the 50 ways refer to company manuals for more detailed information and guidelines.

Waste Reduction Packaging Specifications

The Waste Reduction Action Plan included the Waste Reduction Packaging Specifications. These were based on and expanded upon the "'preferred packaging guidelines' developed by the Coalition of Northeastern Governors' (CONEG) Source Reduction Council."[11] McDonald's continually updates its supplier guidelines on waste reduction in packaging. The company feels it is important for it and other retailers to translate the concern their customers have for the environment to vendors and to provide the impetus for suppliers to produce environmentally friendly products. "The suppliers are a little bit removed," says Bob. "They don't see these customers. And we have an obligation to tell them, 'We care, we want things done; reduce, reuse, recycle.' We want to reduce the packaging itself and the manufacturing effects" of the products they use." Bob calls these "life-cycle assessments."[12]

It's Good Business

A basic premise for being profitable is being efficient, conserving, and being mindful of the economic aspects. McDonald's understands and acts on this premise.

Reduce

A single small packaging change can have extensive impact when rippled through its 10,000 restaurants.

For example, reducing the size of a napkin one inch may save a million dollars. McDonald's shaved only one 16th of an inch off the size of its napkins. It shortened its French fries bags by the same amount. That's not even enough for customers to perceive. Yet with these two changes alone, the company reaped substantial savings in paper costs.

Bob Langert shares several ways to reduce waste in packaging. As guest speaker at the 1994 Portland Recycling Congress, he showed three examples of environmentally friendly Value Meal bags and boxes full of source reduction initiatives. The first bag represented packaging changes introduced to McDonald's 10,000 U.S. restaurants in 1993, which saved the company $775,000. The packaging changes in 1994 saved the company $2.6 million. The "bag of the future" planned for 1995 and beyond is expected to increase those savings even further. But there's more to the bag project than merely cost savings.

Reuse

McDonald's has replaced corrugated trays to hold sandwich buns with reusable plastic trays, and cartons for Coca Cola with large steel canisters. Its meat suppliers have switched from coated cardboard containers to reusable plastic ones. The company has worked with a centralized pool of durable wooden pallets that can be reused. More than half of McDonald's suppliers use pallets from this program, reducing timber demand by 60%. Pump-style condiment dispensers have replaced individual packets in many stores. Every employee at McDonald's USA headquarters has his/her own reusable mug and often one or two guest mugs.

Recycle

At the corporate office in Oak Brook, IL, recycling is as convenient as it can be. Workers separate their food waste from paper waste. They seldom have food at their desks so contaminants are usually not a problem. Beverage stations throughout the facility have signs reminding employees which cups to use for cold drinks, which ones to use for hot drinks, and in which waste containers to dispose of them.

Up to 80% of a restaurant's on-premise waste has the potential of being reduced, reused, or recycled. In the first year of the Waste Reduction Action Plan, the company put in place programs to recycle or to test the recycling and composting of that waste. By 1995, about one-third of the average McDonald's restaurant's waste was being recycled, and close to another 50% of its packaging was being made with recycled materials.

Corrugated shipping containers make up 34% of a typical store's waste by weight. Most corrugated is now recycled, keeping 13 tons of corrugated per store per year out of the landfill and earning in excess of $50 to $100 per ton more than in 1994 in the United States alone.

Closing the Loop

Perhaps the most admirable of McDonald's efforts is its McRecycle USA program established in 1990. As part of McRecycle USA, Ed Rensi made a commitment that, starting in 1991, McDonald's would spend at least $100 million a year buying recycled products for use in its restaurants, including construction materials, trays, booster seats, tiles, safety Playland surfaces, and a large quantity of paper materials. Actually, the corporation spent more than twice that on these recycled products, as well as packaging, supplies, and equipment.

While this may mean lower costs in some areas, this commitment was not based on a cost-saving initiative. Rather, the company was looking at the big picture. As it helps support recycling efforts by closing the loop, it creates the demand in the market place and supports reimbursement rates for recycled materials. By the end of 1994, McRecycle expenditures exceeded $300 million in the United States and almost $100 million abroad.

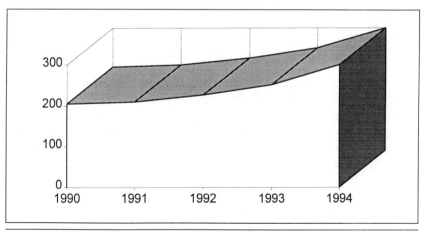

Figure 5-2.   McRecycle dollars spent (in millions).[13]

The corporate office closes the loop, too, in its high quality office paper. Usually the office paper it buys has 50% recycled content with a minimum of 20% post-consumer fiber.

McDonald's initial Waste Reduction Action Plan included switching from white bleached to unbleached recycled carry-out bags. Langert says that when he started working on this project, "Every paper bag supplier in the country said it couldn't be done, and gave me ten reasons why it couldn't." But he didn't give up and finally found a new supplier, Stone Container, who was able to satisfy the company's minimum requirement of 50% post-consumer content. By 1992, all restaurants nationwide were purchasing 100% recycled bags with 50% post-consumer (including newsprint) and 50% pre-consumer content. "Who would ever have thought that we would have been able to switch from white paper bags to brown paper bags that are made from 100% recycled paper?" Happy Meal bags are a minimum of 65% post-consumer waste, largely newsprint. The Big Mac container has 40% pre-consumer and 15% post-consumer content.

Napkins are also made of 100% recycled fiber with 30% post-consumer content. It's true that they have some specks in them. They're no longer bright white because chlorine bleach is not used in the production process. Yet they're sanitary and high quality.

McDonald's has found that recycled products meet both their quality and their economic standards. Bob says that the company has not spent more money as a result of its waste reduction program; it's just spending it more wisely and increasing market demand for recycled goods in the process. Table 5-3 lists the paper-based packaging products used in McDonald's restaurants and their recycled contents.

"Some people think being environmentally responsible means extra cost. But common sense tells us that reducing waste and not polluting the environment in which we work actually reduces costs. That's certainly been our experience from our waste reduction and recycling initiatives—all it takes is a little effort,"[14] says Steve Jermyn, Executive Vice President of McDonald's Australian operations.

Some McDonald's "have walls built of insulated concrete blocks made with recycled photographic film, roofing tiles made from used computer casings, or colorful McDonald's Playland surface made from recycled automobile tires."[15]

Table 5-3.   McDonald's Recycled Packaging[16]

Packaging Item	Total Recycled Content %	Average % PCM[17] content	Average % PIM[18] content	Primary Source of Raw Materials
Big Mac Container	40	15	25	Old corrugated boxes
Carry-Out Bag	100	50	50	Old corrugated boxes
Napkins	100	30	70	Office paper
4-Hole Drink Tray	100	23	77	Newspapers
Happy Meal Bags	65	65	0	Newspapers
Happy Meal Cartons	100	40	60	Newspapers
Trayliners	100	75	25	Magazines
Sandwich Wraps	20	0	20	Industrial scrap
Jumbo Roll Tissue	100	40	60	Office paper
Toilet Tissue	100	30	70	Office paper
Roll Towels	100	30	70	Office paper
Corrugated	40	15	25	Old corrugated boxes

Sharing Knowledge and Resources

"McDonald's shares the McRecycle USA Registry data base with other businesses, trade associations, and government agencies. This has led to the creation of additional 'buy recycled' programs throughout the country."[19] Manufacturers with recycled products to sell can call toll-free for details. They will receive information on the program and how to register their products for McDonald's consideration.[20]

Striving for Excellence

Most projects in the Waste Reduction Action Plan have been completed. Many of them have been implemented and have resulted in more efficient use of resources, lower costs, increased demand for recycled goods, and an improved public image. Some of the more significant projects are summarized in Table 5-4.

Knowing When to Say "No"

But not all projects were successful.

There were attempts to replace plastic stirrers with wooden ones. Results from test markets were positive with a 38% source reduction and

Table 5-4.   Waste Reduction Action Plan Accomplishments[21]

Targeted Area	Action Taken
Polystyrene foam containers	Switched to wax wraps, insulated wraps, and sandwich collars. Resulted in reducing container volume by up to 90% and related corrugated usage by 85%, curtailing amounts of waste going to landfills.
Big Mac wrap/box	Switched from box to 65% recycled (15% PCM) paper wrap. Replaced paper wrap with lightweight corrugated box, 40% recycled content (inner layer), for 10% source reduction.
Sandwich wraps	Converted all insulated sandwich wraps to elemental chlorine free (ECF) paper eliminating dioxins emitted in manufacturing process. Switched from 15# to 14# paper for four sandwich items, reducing paper usage by 1 million pounds per yr.
Quarter Pounder wrap	Reduced size of wrap 1", 8%.
Chicken McNugget box	Redesigned to reduce amount of paperboard 20%.
Happy Meal bags	Redesigned to reduce size by 20%.
	Reduced basis weight from 41# to 30# paper lowering paper usage by 45 pounds per store per yr. Converted to 65% recycled content/PCM.
Hash brown bags	Replaced cartons for 77% source reduction, saving 3.4 million pounds of corrugated per year.
QPC mini-flute	Increased recycled packaging demand by 11 million pounds per year.
Happy Meal box	Redesigned to reduce size by 25%. 100% recycled content, 40% PCM, 60% PIM.
Raw French fry	Reduced basis weight in product packaging for 42%.
French fry box	Redesigned to reduce paperboard usage by 13%.
Napkins	Increased recycled content to 100%.
	Reduced size by 8%. Reduced embossment saving 30 pounds per store per year in case pack cube.
Carry-out bags	Converted to 100% unbleached recycled bags; 50% PCM (including 20% newsprint), 50% PIM Happy Meal bags with 65% PCM (largely newsprint)
Prep pan liners	Switched to newsprint sheets which use groundwood pulp yielding higher pulp utilization and less chemical bleaching.
Ink	Suppliers increased use of soy-based inks in McNugget packaging and all Happy Meal packaging.
Consumer packaging (trayliners, bags, etc.)	Labeled packaging with recycled contents, receiving positive consumer response.
Condiments containers	Switched from individual packets to bulk dispensers for in-store use, reducing volume of corrugated shipping containers.
Corrugated meat shipping boxes	All meat suppliers phased out wax coated boxes in favor of poly-coated recyclable box or bag-in-the-box (box is recyclable corrugated).

Table 5-4.   cont.

Targeted Area	Action Taken
Corrugated shipping boxes for meat and poultry delivery	Replace with reusable plastic containers.
Corrugated cardboard	65% of U.S. restaurants recycling corrugated. Most food suppliers and all packaging suppliers switched to 35% recycled corrugated with minimum 10% PCM. Reduced corrugated needed for hot cups from 175# to 150#. Changed case pack for cold cups and lids saving 160,000 pounds of packaging per year (18 pounds per store per year). Downsized coffee stirrer boxes by reducing head space created by new bulk packaging. Redesigned and reduced cake cone boxes.
Generic containers	Reduced size saving 16.3% in packaging materials. 1994 Earth Effort Packaging Award winner Westvaco incorporated 50% recycled content with 30% PCM.
Side order generic container	Introduced in 1995 as alternative to standard container for small items, saving 450,000 pounds of packaging materials per year.
Cleaning supplies	Eliminated inner-pack dividers in shipping containers. Switched from single portion packets to bulk containers.
Shipping pallets	Since June 1991, more than 50% of suppliers participate in CHEP USA program involving durable wood pallets and centralized pool program, reducing timber demand by 60%
McRecycle USA	In first five years of program, spent more than $1.2 billion ($22,000 average per restaurant per year), doubling the goal. Registered more than 700 companies in program. Shared data base with over 250 outside organizations. Utilized more than 50 new suppliers. Annual awards to outstanding suppliers and McDonald's regional office.
Earth Effort Packaging Award	Recognizes significant environmental achievement from suppliers. 1994 winners: Traex, Georgia Pacific, and Westvaco.
Biannual owner/operator convention	Recycled and/or composted virtually all waste attended by more than 10,000 people for one week including paper, corrugated, glass, polystyrene, and food waste.

high consumer acceptance. The 75% increase in costs, however, made it prohibitive.

McDonald's has also been unable to source totally chlorine-free coffee filter paper, either oxygen bleached or unbleached, since 1991.

Maintaining Momentum

While some people believe that employees can work around management, Bob believes that only with management support and leadership can such programs be truly successful.

With Employees at Headquarters

While the focus is on the restaurants, employees at the corporate office also play a part in protecting the environment. Office workers are encouraged to make two-sided copies, which the copiers track. Disposable cups are recycled as noted earlier. All the computers they buy are Energy Star certified. They automatically power down when not in use for a specified time. Of course, employees are still expected to turn off their systems and other electronic equipment when they go home. The company saves on energy costs, and utility companies reduce pollutants from electricity generation. Articles in the company's weekly newspaper maintain employee awareness.

With Field Operations

To keep the program in the forefront and stimulate teamwork, McRecycle presents a special award every year to its restaurant operations. This distinction goes to the region with the highest percentage of recycled product expenditures in proportion to total construction, equipment, and remodeling purchases.[22]

With Suppliers

Every fall, McDonald's hosts its suppliers conference on the environment. This forum of corporate representatives and vendors reports on quality, pricing, and environmental performance. Attendees develop their business plans for the coming year, and McDonald's presents its annual Earth Effort Packaging Awards. These are bestowed on the suppliers who have excelled in accomplishing their environmentally friendly packaging goals for that year.

With Other Leading Organizations

McDonald's believes in leading by example. In September 1994, it became one of seven members on a task force responsible for researching more environmentally preferable papers under the auspices of the Environmental Defense Fund (EDF). Time, Prudential, Nations Bank, Duke University, and Johnson & Johnson are also on the Paper Task Force. Members are researching ways to incorporate benign bleaching practices and forest conservation. This 18-month project will determine without emotion or bias the paper products they will use throughout their operations—from lavatory supplies to printing paper, computer greenbar, and even packaging.

The task force will define "environmentally preferable paper" through research and collaborative efforts. In the first months of the task force they visited over three dozen paper mills, recycling centers, and related facilities.

Personal Perspectives

Leadership Starts at the Top[23]

Environmental issues never go away. They just take turns being in the spotlight. For a while it may be conservation, topsoil erosion, or toxic waste. Then attention may be turned to landfills, incineration, or recycling. Regardless of what the hot issues are, McDonald's accepts its responsibilities.

According to Shelby Yastrow, Sr. Vice President, General Council and Chief Environmental Officer, the company stopped using polystyrene food containers not because of legislation, but because customers and the public demanded it. "I see things more clearly and have become a true believer in the market place. It works. If we do something wrong, our competitors, customers, and the media will play on it. Legislation is not the way. The law of the market place will force the consumer-driven company to do the right thing."

McDonald's has encouraged other companies to follow its example. Georgia Pacific produced McDonald's first 100% recycled newsprint trayliners. These trayliners deliver the environmental messages printed on them to over 10 million customers every day, increasing awareness and providing solutions.

"I'm proud of what we've accomplished. We've won the President's award and many other awards. Our customers wanted it, and we gave them what they asked for." McDonald's doesn't keep its environmental innovations to itself. "The day I made the announcement about McRecycle, I invited our competitors to use the database."

Shelby takes his environmental values home with him and shares them with his family. "I've become kind of a pest at home and in the office. My family recycles everything. At our (vacation) home in Scottsdale, AZ, we separate our trash. Then we take it ourselves to a recycling center in Phoenix because Scottsdale doesn't have curbside recycling." Like Bob Langert, his family does their own composting. He practices energy conservation, too, "walking around the house at night turning off lights." In dry, hot weather, his sprinkler system helps him conserve water. You won't catch him causing unnecessary pollution by idling his car to warm it up in the winter or to air condition it in the summer. When shopping, Shelby looks for recycled content in packaging and even in greeting cards. While the dilemma of "paper or plastic?" rages on, the Yastrows choose plastic because they bring them back to the supermarket for recycling. "In the office my scratch pads have writing on both sides. It kills me to see someone take a full sheet of paper just to write a phone number."

What are his pet peeves? One is excess packaging. "For a couple of ounces of perfume, there's this big bottle wrapped in tissue and cardboard sitting in a plastic sleeve, then boxed and shrink wrapped." He is very proud of the progress McDonald's and its suppliers have made in reducing packaging waste.

Then there's the newspaper. "I get two newspapers every day and read a little of each." His time is limited so he reads only the news that's most important to him. "It tears my heart to take them out to the trash at night." At least he mitigates the loss by putting them in the recycle pile, and newsprint is made from recycled stock.

Why reduce, reuse, and recycle? "Some do it because it's good for business. Some do it because it's good for the environment." McDonald's has proven that both can be done profitably.

Lifestyle

In the mid 1980s, Bob Langert worked for Perseco, the packaging company that services McDonald's. Charged with eliminating CFC-based packaging, he became involved in environmental issues. Through his involvement in the EDF project, Bob became an integral part of the task force and eventually a full-time McDonald's employee.

Bob's bookshelf is evidence of his commitment to and knowledge of environmental issues. One of his favorite books is *The E-Factor* by Joel Makower because of its positive and innovative outlook on how companies can implement environmentally sound practices into their businesses. At his recommendation, McDonald's Corporation donated copies of *The E-Factor* to 100 universities across the United States. Bob feels that any university offering courses on the environment should have a copy of the book in its library.

What does he do outside the job site? Bob and his family have incorporated environmentally responsible habits in their home for years. They recycle, of course, and use energy-efficient lighting. They eliminate all organic wastes (except for human waste) by vermicomposting. Some estimates put the amount of organic waste that goes into our landfills at 7–8%. By reducing consumption, reusing, recycling, and composting, the Langert family of four generates only two grocery bags full (less than 30 gallons) of garbage for pick-up each week.

The vermicomposting bin looks like another recycling bin, except that it contains dirt and live red worms. The Langerts throw all their organic wastes into the bin, which they keep in the basement. In a day or two, the worms turn everything into rich dirt. Since organic waste is primarily water and moisture evaporates in the composting process, very little new "dirt" is added. It's surprisingly odorless, too, he says, as the worms do their work before the waste has a chance to decay.

Bob anticipates innovations we can't yet foresee. He is "convinced that 50 or 100 years from now, people will look back at our time and think that we were cave men and women, putting our (trash) in the middle of the ground. It just doesn't make any sense."

Hands-On Leadership[24]

Owner/operator Mark Brownstein is in his sixteenth year as a McDonald's franchisee. His father started in the business in 1969 and expanded to four stores. Mark joined him in 1980. Through growth and acquisition, they now own 13 stores in Orange and Los Angeles Counties. During his college years, Mark was active in environmental issues. Although he became an attorney, he eventually decided to change professions so that he could have an influence.

When McDonald's embarked upon its environmental mission, it sought coordinators in each region. Mark volunteered, became Environmental Coordinator for the Los Angeles Region, and went into battle.

A single waste hauler provided over 100 stores with color-coded bags for separating large portions of their waste stream. Within a short amount of time, the region was recycling 70% of its waste stream. About one third of it was corrugated. Another third was food wastes—coffee grounds, broken eggs—which went to feedstock. The program was easy to follow, diverted a lot of waste, and was a little cheaper than sending everything to the landfill.

Mark's stores were recycling 70% of their waste stream long before California's 25% diversion deadline, so he started looking for ways to reach the stores in "franchised" cities.

Many Southern California cities franchise the waste management contracts. That is, they award exclusive waste management contracts, often at the expense of independent haulers. This inhibited McDonald's Los Angeles Region from expanding the program. Stores in franchised cities paid $450 per month to send their trash to the same landfills as those in unfranchised cities, who paid $180 per month. Their waste volumes and profiles were similar (number of bins, days for pick-ups, etc.). But the program fell apart because of economics. Today, recycling consists of only corrugated so that the recovery rate for all the stores in the Los Angeles Region is 33%.

"I'm real proud of what McDonald's has done as a good corporate citizen. The company influenced the market to include recycled content in corrugated. When McDonald's decided to establish McRecycle USA, it distinguished us from our competitors, stimulated the market, and helped create jobs. By deciding to quit using polystyrene, McDonald's has been instrumental in bringing about new technology."

Mark offers simple, straightforward advice to organizations of all sizes.

- Recycle corrugated, even if it means finding a small independent company with just a pickup truck.

- Office building tenants should join forces and negotiate for consolidated pick-ups and the best rates.

- Building managers can offer incentives (free lunches, champagne, rent credits) to tenants who participate in building-wide recycling.

- Close the loop and buy recycled products even if they cost a little more.

- Recycling isn't always the answer—source reduction is.

He reminds management about what's important by distinguishing managing from leading: Managers do things right—leaders do the right things.

Innovative Vision

Bob has used his own experience with composting to bring innovation to McDonald's. The restaurants are limited in what they can recycle because so much of their trash is food waste. In the fifteen Albany, NY, stores, all food and paper wastes generated behind the counters are put into buckets. The waste is then sent to a composting facility where it is mixed with other organic wastes, turned into a rich humus, and sold as fertilizer.

Outside the United States, McDonald's has implemented other waste disposal and recycling techniques. In the Netherlands, for example, the behind-the-counter food wastes are sold to pet food manufacturers to produce cat food.

Recycling the waste that their customers generate is a more difficult challenge. Bob envisions meeting that challenge "down the road." There are some significant hurdles that must first be cleared.

McDonald's Corporation uses paper-based packaging instead of polystyrene for its menu items. As stated earlier, the late 1980s saw a growing controversy about polystyrene versus renewable materials. The company tried recycling its patrons' polystyrene trash. But getting customers to throw their used recyclable food containers into one receptacle and the rest of their waste into another met with only limited success. Educating such a large and diverse customer base is a massive task.

McDonald's envisions that someday its consumer packaging and utensils will be homogeneous with food waste for easier disposal and recycling. Existing alternatives are either not economical or not commercially available. This may require "changing the paradigm," as Bob Langert says, and looking beyond traditional materials to some that perhaps are yet to be created. After all, in the 1930s, who could have predicted that polystyrene would be invented in the next decade and go on to be so widely accepted? Likewise, who's to say what new packaging will be available in the future?

Bob restates two of McDonald's major achievements:

> "established several new programs to ensure our environmental commitment continues. Some of these have already been presented: (1) creating an Optimum Packaging Team where inter-departmental managers regularly review opportunities to reduce and/or eliminate secondary and shipping packaging, and (2) expanding our internal McRecycle Team to continue to find new opportunities of using recycled materials in every area of our business."[25]

The Bottom Line

It's possible to make plenty of progress by doing what's good for both business and the environment.

The key that will make or break the implementation of environmentally preferable alternatives in any organization is whether or not those alternatives are cost competitive. "I think that people are more than willing to do things for the environment as long as it's cost compatible, says Bob Langert. "You find a lot of resistance when it's not."

Reduce, reuse, recycle, close the loop, and involve everyone. This total approach ensures long-term success for McDonald's, its employees, vendors, customers, and the global community. This "across the board commitment to instilling environmental concerns at all levels of the system is good for the environment and good for McDonald's business."[26]

McDONALD'S WASTE REDUCTION POLICY[27]

McDonald's believes it has a special responsibility to protect our environment for future generations. This responsibility is derived from our unique relationship with millions of consumers worldwide—whose quality of life tomorrow will be affected by our stewardship of the environment today. We share their belief that the right to exist in an environment of clean air, clean earth and clean water is fundamental and unwavering.

We realize that in today's world, a business leader must be an environmental leader as well. Hence our determination to analyze every aspect of our business in terms of its impact on the environment, and to take actions beyond what is expected if they hold the prospect of leaving future generations an environmentally sound world. We will lead, both in word and in deed.

Our environmental commitment and behavior are guided by the following principles:

Effectively managing solid waste—We are committed to taking a "total lifecycle" approach to solid waste, examining ways of reducing materials used in production and packaging, as well as diverting as much waste as possible from the solid waste stream. In doing so, we will follow three courses of action: reduce, reuse and recycle.

Reduce—We will take steps to reduce the weight and/or volume of packaging we use. This may mean eliminating packaging, adopting thinner and lighter packaging, changing manufacturing and distribution systems, adopting new technologies or using alternative materials. We will continually search for materials that are environmentally preferable.

Reuse—We will implement reusable materials whenever feasible within our operations and distribution systems as long as they do not compromise our safety and sanitation standards, customer service and expectations, nor are offset by other environmental or safety concerns.

Figure 5-3.   McDonald's waste reduction policy.

Recycle—We are committed to the maximum use of recycled materials in the construction, equipping and operations of our restaurants. We are already the largest user of recycled paper in our industry, applying it to such items as tray liners, Happy Meal boxes, carry out bags, carry out trays and napkins. Through our "McRecycle" program, we maintain the industry's largest repository of information on recycling suppliers, and will spend a minimum of $100 million a year on the use of recycled materials of all kinds. We are also committed to recycling and/or composting as much of our solid waste as possible, including such materials as corrugated paper, polyethylene film and paper. We will change the composition of our packaging, where feasible, to enhance recyclability or compostability.

Conserving and protecting natural resources—We will continue to take aggressive measures to minimize energy and other resource consumption through increased efficiency and conservation. We will not permit the destruction of rainforests for our beef supply. This policy is strictly enforced and closely monitored.

Encouraging environmental values and practices—Given our close relationship with local communities around the world, we believe we have an obligation to promote sound environmental practices by providing educational materials in our restaurants and working with teachers in the schools.

We intend to continue to work in partnership with our suppliers in the pursuit of these policies. Our suppliers will be held accountable for achieving mutually established waste reduction goals, as well as continuously pursuing sound production practices which minimize environmental impact. Compliance with these policies will receive consideration with other business criteria in evaluating both current and potential McDonald's suppliers.

Ensuring accountability procedures—We understand that a commitment to a strong environmental policy begins with leadership at the top of an organization. Therefore, our environmental affairs officer will be given broad-based responsibility to ensure adherence to the environmental principles throughout our system. This officer will report to the board of directors on a regular basis regarding progress made toward specific environmental initiatives.

On all of the above, we are committed to timely, honest and forthright communications with our customers, shareholders, suppliers and employees. And we will continue to seek the counsel of experts in the environmental field. By maintaining a productive, ongoing dialogue with all of these stakeholders, we will learn from them and move ever closer to doing all we can, the best we can, to preserve and protect the environment.

Figure 5-3.   cont.

ENDNOTES

1. *Waste Reduction Task Force—Executive Summary* (New York, NY, NY: Environmental Defense Fund and McDonald's Corporation, April 1991), p. 4.
2. *Waste Reduction Task Force—Final Report* (New York, NY, NY: Environmental Defense Fund and McDonald's Corporation, April 1991), p. 31.
3. *Facts & Resources...in a small package...McDonald's Earth Effort* (Oak Brook, IL: McDonald's Corporation, 1994).
4. Personal interview with Bob Langert, October 1994.
5. *Environmental Affairs Newsletter* (Oak Brook, IL: McDonald's Corporation, January, 1992), p. 4.
6. *McDonald's Earth Effort: Our Commitment to the Environment, Our Resource Conservation and Energy Policy* (Oak Brook, IL: McDonald's Corporation, 1994).
7. *The TriAce Vision* (Hinsdale, IL: TriAce).
8. *Waste Reduction Task Force—Executive Summary*, p. 4.
9. *Facts & Resources...in a small package...McDonald's Earth Effort.*
10. *News From McDonald's Corporation: Calling All Phone Books Recycling Program to Begin in East/South County* (San Diego, CA: Stoorza, Ziegaus & Metzger, Inc., August 17, 1992).
11. *Waste Reduction Task Force—Executive Summary*, p. 10.
12. Personal interview with Bob Langert, October 1994.
13. Internal memo from Robert L. Langert, Director of Environmental Affairs, McDonald's Corporation, June 30, 1995, p. 1.
14. *McDonald's Earth Effort Global Express* (Oak Brook, IL: McDonald's Corporation, June 6, 1994), p .2.
15. *The Planet We Share* (Oak Brook, IL: McDonald's Corporation, 1992).
16. *McDonald's Waste Reduction Action Plan: Status Report 1995 Goals* (Oak Brook, IL: McDonald's Corporation, June 30, 1995), p. 9.
17. Post-Consumer Recycled Material—waste material generated from products or packaging which have served their intended use and have been recovered, reprocessed and reused as a raw material for the manufacture of another marketable product, according to the *Waste Reduction Task Force—Final Report*, p. 80.
18. Post-Industrial Material (pre-consumer)—manufacturing waste generated during the intermediate steps of producing an end product, but excluding materials (such as mill broke) that are routinely internally recycled to make the same, or a very similar product, according to the *Waste Reduction Task Force—Final Report*, p. 80.
19. *McRecycle USA* (Oak Brook, IL: McDonald's Corporation, 1992).
20. Ibid.
21. *McDonald's Waste Reduction Action Plan: Status Report 1995 Goal.*
22. *Environmental Affairs Newsletter* (Oak Brook, IL: McDonald's Corporation, February 10, 1992), p. 2.
23. Personal interview with Shelby Yastrow, August 1995.
24. Personal interview with Mark Brownstein, August 1995.
25. Personal interview with Bob Langert, October 1994.
26. Ibid.
27. *Waste Reduction Task Force—Executive Summary*, pp. 17–18.

CHAPTER 6

MERRILL LYNCH, WORLD FINANCIAL CENTER, NY

Background

Merrill Lynch has been recycling for over 40 years, starting with the computer punch cards used for data entry of employee and client records. Although the punch cards were being recycled, the white paper printed reports were not, until 1980.

Now Vice President of Facilities under the Corporate Real Estate organization, Andy Lauro started with Merrill Lynch in 1966. In 1980, he was property manager for the company's New York City facilities. Back then, there were six to seven thousand employees occupying 50 floors at 165 Broadway. Andy would walk the sites regularly. He was struck by the large amounts of paper being trashed and recognized the potential for recycling it. He also wanted to consolidate and enhance the many individual recycling efforts throughout the company.

At that time, each department was like a company within a company, responsible for its own operations. It was not unusual for the keypunch department to make $10,000 a year on recyclables.

Andy knew he needed management support to be successful. He got it by sending his message up the ladder, presenting his business case to his department manager, who then took it to his director, who in turn got the Vice President behind the concept.

In 1981, top management authorized Andy to take responsibility for all of the company's recycling efforts in New York City. This included both hauling and reimbursements.

Initially the departments weren't at all happy about this decision to centralize control—and revenues. Andy met with the four or five departments that were generating computer waste paper and explained how recycling revenues would offset occupancy costs. Their budgets (and every department's budget) were already being charged for the space they occupied. The money saved would help reduce those charges not just for them, but across the board so that everyone would benefit. He finally got them to buy into the basic concept.

The next year, Andy put his plan in motion. He contacted the Council on the Environment of New York City to help him implement a white paper recycling program company-wide.

CENYC

"The Council on the Environment of New York City (CENYC) was formed in 1970 and is a privately funded citizens' organization in the office of the Mayor. The CENYC promotes environmental awareness among New Yorkers and develops solutions to environmental problems."[1] Ann Marie Alonso, Director of Waste Prevention & Recycling Services, works with businesses throughout the city to carry out the Council's charter. She and other members of CENYC help the community protect the environment by showing businesses how to comply with waste and recycling regulations.

The Council helped Andy by first conducting a waste audit and determining the paper flow of Merrill Lynch's headquarters facilities. It identified the waste paper generators, estimated the revenue potential, and calculated the costs that could be avoided. The CENYC also held brief educational sessions for the Merrill Lynch employees who would be affected. Andy wasn't very knowledgeable about recycling or how to set up an effective program, so the Council did it for him. The resulting program was focused on computer-related waste paper, both white ledger and greenbar (CPO), and served as the model for the next few years.

Implementing the Plan

Computer technology was just starting to catch on in corporate America and was picking up steam within Merrill Lynch. Therefore, the plan focused initially on only the computerized areas of the company. Andy recalls, "These little computer rooms were sprouting up all over the place. As a computer room was developed, which comes out of our group here on the facilities side, I would go in and implement a recycling program right in that room."

His first full year of consolidated recycling efforts generated $80,000 in revenue, an impressive amount in 1982. He knows that their trash volume went down, but back then the janitorial service and the trash hauler didn't track cost avoidance.

With plans of relocating elsewhere in the city, the company scaled back further development of the recycling program for a couple of years. Merrill Lynch finally moved its headquarters to the World Financial Center in 1985. Andy's department used the move to expand the recycling program beyond high grade white paper and to work areas outside the computer rooms. Within the first three months, about twenty employees called him complaining that it was hard to comply with the program and that it was inconvenient. Andy won them over by explaining that he needed every employee to participate to make this management-backed program a success.

In 1986, trash hauling rates in the city more than doubled from the previous year, although Merrill Lynch's rates stayed the same under its contract. Revenues held steady and cooperation was good. Then in 1988, Andy learned that the Department of Consumer Affairs had authorized the trash haulers to increase their hauling rates to the maximum allowed in their agreements. Just one year later, the company's trash hauling costs had risen 110%, taking a tremendous bite out of the budget. Also, the City of New York was discussing mandatory recycling starting in 1990.

The company decided to be proactive and expand the existing program. "We felt that with the increase of trash hauling costs and the city's plans to mandate recycling," says Andy, "why not pioneer this and do it now? Why wait until the law comes into effect? Let's go with a full desktop recycling program."

So in October of 1989, the Council of the Environment came back and surveyed one of Merrill Lynch's main sites housing several thousand employees. The CENYC went through office by office, desk by desk, researching the type and quantity of waste generated and presented the findings to Andy. To kick off the mixed paper recycling program, Merrill Lynch informed and educated everyone from the top down by corresponding first with senior management, then with middle management, and finally with supervisors and general staff. Announcements and flyers explained the forthcoming program. A class schedule was posted in the cafeteria.

Responsibility for recycling rested largely within the Corporate Real Estate group, which was instrumental in getting the program off the ground.[2] Coincidentally, Andy had been getting a sudden flurry of calls from employees asking why they weren't recycling. "It's coming," Andy told them. His timing couldn't have been better. Andy went with a single can program because he thought people wouldn't take the time to separate white ledger from mixed paper. He relied a lot on his janitorial team.

Brochures and leaflets alerted employees about the forthcoming program. The company set up 15-minute slide shows in the cafeteria and ran them continuously throughout the day. The slide shows were short and easy to view during breaks and lunch time.

The Janitorial Manager had responsibility for cleaning all Merrill Lynch facilities in Manhattan and northern New Jersey. He helped educate secretaries, managers, his crew, even himself. "We've come a long way," he said. "I never told the people about cost avoidance even though it costs me less to have that garbage removed because at that time, they didn't care. They had more important issues on their minds such as no raises and frozen bonuses. And here we were, asking them to give even more to the company. We told them to start recycling their notes and memos." To make it easy, every employee received a separate container for waste paper. There were also larger bins in convenient locations throughout the buildings.

Questionnaires were sent to 6,000 employees asking them what they thought of the program and what else Andy Lauro and his staff could do to make it better. Andy remembers the results of that survey. Of the 60% who responded, 80% felt the company could do more.

"Some said, 'This stinks' and 'It's a waste of time.' Others were really into it, asking about recycling other things." Only two people said that someone should come to them and pick up their waste paper.

The survey also stimulated a lot of interest. Many New York employees lived in New Jersey where recycling had been mandatory for several years, so they were used to it. Andy received many positive calls from employees asking what else the company was doing to support recycling. It seemed that all the employees had been waiting for it and were happy to participate.

"We put the program into play and waited a month. Then the janitorial service monitored the program by looking into waste barrels to pinpoint trouble spots." Two weeks after this first audit, the Janitorial Manager conducted another one, floor by floor. There was still a big gap between what the company wanted and what was actually happening. So Andy went throughout the company and talked with managers about the importance of making the program successful. They, in turn, had meetings with their people, reiterated the goals and guidelines, and motivated their people to cooperate.

Employees incorporated recycling into their daily routines, emptying their personal bins into the larger containers on the way to lunch, the coffee machine, and the lavatory. Because the janitorial staff didn't empty the desktop containers, there were no additional labor costs or union issues.

The Janitorial Manager made sure everything was being picked up and staged properly. Andy's overall staff and five Building Managers carried out and reinforced the program in all sites with floor inspections. Andy sent memos reminding office workers about the program and telling them how many trees they had saved. "Slowly but surely, people would do more. Then they started asking about recycling newspapers and magazines. Now we really don't run into any problems in New York."

That became their pilot program after which the other sites' recycling programs were modeled. It caught on like wildfire. In early 1990, the expanded program was implemented throughout the company's New York City offices. Later that year, it was replicated in the central computer site. By December 31, four of their five New York facilities were on board. Merrill Lynch reduced its garbage container pulls in 1990 by 80% of its 1989 volume: from two per day to one every three days (from 10 down to 2 per week).[3]

The fifth site with 2,000 operations people was scheduled to relocate in early 1992. They decided to stay with the high grade white paper recycling until they moved to their new location in Jersey City, NJ. "As they walked in the door," says Andy, "they received building policies and a description of what they had to do on their desks regarding desktop recycling. So that program went into place as they moved in."

Staying on Track

Andy sends his building managers monthly revenue updates. He also publishes progress reports periodically in the company newsletter, telling everyone how many trees, barrels of oil, and other precious resources they helped save. "You have to keep reiterating the program. At least once a year, you have to go back and tell the people (that) we're still recycling," especially since people move around within the company. During monthly inspections, Andy's people look for work areas where the desktop receptacles are missing, and they replace them. "Reinforcement is very important. Public relations is very important."

Andy handles all the company's New York facilities with about 10,000 employees. He tracks cost avoidance as well as recycling income. He also works closely with their managing agent at the Jersey City site.

Andy has been a very active committee member of the National Paper Recycling Coalition in Washington, DC. He has a personal interest in keeping the recycling program going strong. Raised by parents who struggled through the Great Depression, he learned to never waste. He Andy sees his firm as a type of paper factory. He has focused his efforts on paper because it is the largest single component of the waste stream. Opportunities for recycling other materials such as glass, aluminum, and plastics exist. However, paper is where he can have the biggest impact.

Andy has fulfilled his original dream of maximizing revenues and minimizing waste for the company. With cost avoidance, the total net income shown in Table 6-1 has remained fairly constant.

Table 6-1.   Merrill Lynch Paper Recycling and Cost Avoidance Results for New York City Headquarters Facilities, 1990–1994 (Source: Merril Lynch)

	1990	1991	1992	1993	1994	5-Year Total
Total Tonnage	1,294	1,118	1,045	1,466	1,769	6,692
Total Contribution to the Bottom Line4	$506,035	$355,258	$354,360	$369,858	$391,115	$1,976,626
Contribution per Employee (approx. 10,000 employees)	$51	$36	$35	$37	$39	$198

The numbers for individual divisions changed from year to year because employees moved between facilities. The volume of paper used and cost avoidance figures have fluctuated, partly because until the middle of 1990, Merrill Lynch maintained an internal print shop in New York. Tritech, the company's printing division in New Jersey, now handles all large printing jobs. Tritech participates in Merrill Lynch's New Jersey program under Michael DeNardo, Andy's counterpart for facilities in New Jersey, Florida, and Colorado.

Across the River and Across the Country

The Jersey City laws are different. Waste generators are required to keep food waste (wet waste) out of the recyclables and trash. When workers didn't comply and wet rubbish did creep in, Merrill Lynch was fined $200 by the landlord, who incurred additional cost. According to the Janitorial Manager, "It really affected the (success of) desktop recycling there. We went from 80 cans of (wet) food waste to 11, and then to only 4 cans"—an amazing reduction of 95%!

Seeing Double

Based in New Jersey, Michael DeNardo manages the company's New Jersey, Florida, and Colorado facilities. The hauler for the New

Jersey locations picks up and sorts waste paper for the mills to refabricate into corrugated cardboard and sell at competitive prices.

Mike has implemented the hauler's two-can system. Every employee has two baskets—one for mixed paper, one for trash. Additional containers for newspapers, aluminum cans, and glass are centrally located on each floor. The hauler donates the proceeds from glass and cans on behalf of his customers to the St. Barnabus Burn Center (see Chapter 4). Table 6-2 shows the volume of recycled paper for a five-year period.

Table 6-2.   Volume of Recycled Paper for Merrill Lynch, New Jersey Facilities 1990–1995 (Source: Merrill Lynch)

	1990	1991	1992	1993	1994	1995*	6-Year Total*	Annual Average
Total Tonnage	555	505	495	580	478	544	3,157	526
Pounds per Employee per Year	686	570	492	515	391	351	3,005	501

*projected

What's his secret formula for success? People. "People take the time, people care." Some evening workers saw the janitorial service co-mingling everything in a single bin and called Mike to report improper disposal. These conscientious employees were relieved to hear that the bins had dividers in them. The janitors were putting recyclables on one side and trash on the other. "People even bring in their junk mail for recycling," says Mike, and he does, too.

Overcoming Hurdles

People weren't always this cooperative. In 1988 when the program began in the New Jersey facilities, employees thought it would be hard to do. They complained in anticipation. Mike sent a memo to managers asking for one representative from each department to be on his committee. At the first committee meeting, he laid out the plan and enrolled everyone in the mission. They communicated the company's goals to their respective co-workers so that everyone became part of the solution.

Mike then distributed the individual bins with a memo to every employee. He installed a Recycling Hotline for people to call with questions or problems. Winning over the skeptics was easier than he thought. In five years, only one call came into the Recycling Hotline, an indication of the employees' full cooperation. "They were ready for it. They're used to residential recycling in New Jersey, so they already had the right mindset."

Tritech is Mike's biggest success. This multi-functional division services Merrill Lynch and other clients nationwide. Tritech handles mail distribution, statement printing, and commercial printing such as newsletters, financial reports, and proxy services.

In the Trenches

Mike credits Tritech's Corporate Real Estate Site Manager for contributing substantial dollars to the company's 1994 recycling revenues. Most of Tritech's output is high grade white ledger. The Site Manager makes sure that it's separated from the mixed paper so that the company earns the highest reimbursement rates possible.

In 1989, he was losing money. The hauler he was then using would pick up his computer paper without charging a hauling fee—and without paying him. His present hauler, on the other hand, pays him for his waste paper and reduces his hauling costs, which are around $1,800 per month. He believes his monthly hauling costs would be five times higher. With very little storage space available, he relies on his hauler to come when his bins are full.

The Colorado facilities remain a challenge for Mike. Denver management requires the haulers to shred all documents on the company premises. Instead of making money, Merrill Lynch must pay the trash haulers extra for doing the shredding. Mike's plan to resolve this is to contract a bonded recycling company that will guarantee confidentiality and proper destruction. Then all his janitorial service has to do is to put the paper into a compactor, unshredded. The contractor will shred, haul, and manage the recycling of all of it for him just as his hauler in New Jersey does.

An even bigger challenge is in Merrill Lynch's Florida locations where success has been limited. With Mike's perspective, resourcefulness, and history of achievement, it's just a matter of time before the company's Florida facilities become role models for their neighbors.

Community Champions

Credit for much of Merrill Lynch's success goes to people outside the company. Who are they? How did they contribute to success?

Earth Day New York

Merrill Lynch's Manager of Printing Services says that Pamela Lippe was particularly instrumental in educating "everyone in the market" on closed loop recycling. As Executive Director of Earth Day New York, Pamela oversees city and chairs activities such as workshops and seminars Earth Day and all year long that promote conservation, recycling, and various aspects of creating a sustainable economy. She works with organizations of all types and sizes throughout the city to promote environmentally friendly policies, attitudes, and habits.

The 1995 Earth Day conference agenda focuses on developing a sustainable economy in the United States and around the world. In 1993, the Earth Day Closed Loop Recycling Conference used corporate success stories to show participants how closed loop recycling works. It explored the business and economic aspects, legislation, and ways to create markets for recycled "waste." This excerpt from the Program guide describes the conference's mission:

> The Earth Day Closed Loop Cooperative is an innovative new program that helps corporations and other institutions coordinate their buying power and waste materials in a way that builds the markets for recycled materials and can save the purchaser money.
>
> The Earth Day Closed Loop Cooperative will help to:
>
> • Protect the environment.
>
> • Drive prices for recycled products down rather than up (making them more affordable).

- Establish new markets and broaden the customer base for recycled products.

- Create a steady stream of reusable material that can stabilize prices and supply.

- Educate corporate and governmental decision-makers about how recycling makes good business sense *and* can help save the environment.

- Generate American jobs and attract new industry into the region.

- Provide added value to participating corporations in the key areas of community responsibility and public relations.

Recycling is an economic development issue for business—just as energy conservation was ten years ago. Corporate leaders with vision have been actively pursuing this approach for years. They understand the advantages not only to society but to their own companies. This conference will present corporate success stories and introduce you to Closed Loop Recycling.[5]

The program was based on the success at Home Box Office. Just ask Pamela or people at Merrill Lynch about closed loop recycling, and they all have the same response: "You should talk to Steve Laughlin at HBO."

Good Neighbor

All of HBO's corporate letterhead and envelopes are 25% cotton plus 37% recycled content from the firm's own waste paper. Its copy paper is also recycled, although not from its own paper waste. Stephen Laughlin is largely responsible for the use of so much recycled fiber.

Steve became Environmental Affairs officer at HBO/Time Warner at the end of 1994. He was one of the pioneers in closing the loop while serving as a purchasing agent in the company's Facilities Department. Working closely with haulers, recyclers, and mills to solve the waste paper dilemma, he saved his company over $40,000 in 1992. He never did it with the sole purpose of saving money, but as part of an overall strategy to use resources wisely.

He didn't view recycling as a loop either, but as a three dimensional diamond. "You use the facets to build relationships and possibilities. Usually when you're responding to a market of scarcity, the more you buy, the more you drive the price up. But in a closed loop, the more you buy, the more you drive the price down because you're supplying more raw materials to make production cheaper for manufacturers. We will supply the post-consumer portion of this material. We are looking for a manufacturer to participate and be the 'best deal in town,' because he/she will secure these customers (like Merrill Lynch) for future business."

Steve does business with paper manufacturers that have plants in the Northeast, such as International Paper's Springhill Recycled Relay DP factory in Ticonderoga, NY. Byron Western, a division of Crane's Company, also has plants in the area. By doing business with regional companies, HBO contributes to the tax base, the local economy, and the job market.

He believes that private industry is better positioned to drive closed loop recycling than government. That's because most government agencies must go through RFPs (requests for proposals) that take a long time, are very complex, and must go to the lowest bidder.

"My responsibility to the environment is to achieve zero waste out of headquarters. Part of that is to make sure that my waste is being remanufactured and not going to the landfill, and that my waste hauler understands that he/she gets not just my business but also my commitment. I can help my hauler find new markets."

Steve also helps other commercial waste generators in New York with their disposal dilemmas. His experiences implementing HBO's waste reduction and recycling program and his network of contacts have polished other facets of the recycling diamond by contributing to his neighbors' successes.

Corporate and Personal Responsibility

Internal Leadership

Steve Laughlin isn't the only good neighbor. Merrill Lynch employees return the favor by donating aluminum cans to We Can. This nonprofit organization provides basic medical and dental attention, housing, and employment to the homeless.

Several Merrill Lynch employees involved in the company's recycling and closed loop efforts help other New York organizations with their programs by sharing their experiences. Andy Lauro is often a guest speaker at other investment companies such as Prudential-Bache and Paine Webber, telling them about Merrill Lynch's successful program.

Bullish on People and the Environment

Andy doesn't have any personnel problems. Average longevity on his staff is 15 years with very little turnover. He believes in promoting from within the company. Four of his five building managers came up through the ranks. He himself worked his way up from building manager for this and another facility to his current position, which he has held since 1989. Andy feels that the building operations and facilities industry has been given a "bad rap" in the past. "We're part of the overall team. Whatever happens out there happens here," and vice versa. He fosters a family atmosphere and team spirit in his people.

He has always conserved, whether at work or at his home in Queens. This New York borough has provided bins to all of its residents. Andy has made it a habit to separate newspapers, throw food and wet waste down the garbage shoot, and even rinse cans before placing them in the recycle bin.

One of Andy's stars was his janitorial supervisor. According to Andy, she saw to it that her entire crew was properly trained on what to throw out, making sure plastic bags didn't get into the mainstream. She continually followed up to make sure that there were no contaminants in Merrill Lynch's compacted trash and that it was clean, paper-based rubbish containing no furniture, air filters, old office supplies, etc. "Our waste hauler doesn't pay us for it (compacted clean rubbish)," Andy continues, "but doing this does reduce our hauling costs." The cost avoidance figures in Table 6-1 take into account the lower tonnage of waste paper as well as the reduced volume. Andy says of the waste paper market, "It's like the stock market—up today, down tomorrow." The paper reimbursement rates in Chapter 10 illustrate his point.

A Team Effort

Howard Sorgen, Senior Vice President for the company's Information Services, has also been a key player in Merrill Lynch's formula for success. Under his leadership, members of the Print Reduction Committee contributed many inventive ideas, such as promoting two-sided printing, completely cancelling unnecessary publications, and eliminating or reducing the distribution of certain reports.

The paper savings were staggering. In 1991, more than 616 million pages were eliminated from the print queue.[6]

Every employee has to participate in recycling for two reasons. First, it's mandated by New York City law. Second, it's the responsibility of each Merrill Lynch employee to carry out the company's mission and adopt its philosophy.

View from the Top

When Daniel P. Tully became CEO, he published five principles that have become the cornerstone of the company's philosophy. These five principles have been printed (on recycled paper, of course) and distributed to all employees. New employees receive them in a complete orientation package from Human Resources. The package includes information on the company's recycling program as well.

Here are the Merrill Lynch Principles:

- **Client Focus**: The client is the driving force behind what we do.

- **Respect for the Individual**: We respect the dignity of each individual, whether an employee, shareholder, client or member of the general public.

- **Teamwork:** We strive for seamless integration of services. In the client's eyes, there is only one Merrill Lynch.

- **Responsible Citizenship**: We seek to improve the quality of life in the communities where our employees live and work.

- **Integrity:** No one's personal bottom line is more important than the reputation of our firm.

Closing the Loop

With the efforts of the Assistant Vice President/Section Manager of Facilities Purchasing, the company closes the loop by buying recycled goods such as paper products made exclusively from recycled fibers. Use of some environmentally unfriendly products such as yellow legal pads has been discouraged.

The haulers and mills work with many of their large clients like Merrill Lynch to provide paper goods manufactured from their clients' own recycled paper. Because of the size of these companies and the amount of waste paper generated, the mills can isolate their production runs. They process each client's own waste paper separately, and sell it back to the client for its exclusive use. The resulting products include printing paper, paper towels, napkins, and boxes for company cafeterias.

The company's Purchasing Department works with its waste haulers and the paper mills to recycle and buy back their waste paper. "We first became involved in desktop recycling in 1988 and closed loop recycling in 1992," says this Assistant Vice President. "As soon as you say 'printed from your company's own recycled paper,' people ask you to prove it. Sometimes you have to follow your paper to the mill and through the process." The Vice President of Purchasing advises, "You have to investigate and educate yourself to really save anything."

Merrill Lynch uses its own recycled paper for special projects like the annual report, which the Printing Services Department requested to be environmentally responsible. By using its own recycled paper for the financial section, Merrill Lynch saved a lot of natural resources. It does not do this for general use because it gets too onerous for the company to track and for the mills to store until the manufacturing run.

"Our use of recycled paper has jumped dramatically," says the Manager of Printing Services. A major portion of Merrill Lynch's marketing literature is printed on recycled paper. The company also uses recycled paper for most of its monthly publications, inserts in customer statements, research materials, and some of the company stationery.

Table 6-3.   Natural Resources Saved 1994

Paper Recycled (tons)	Trees	Oil (barrels)	Water (gallons)	Electricity (kilowatts)	Air Pollution (pounds)
3,000 tons	51,000	6,000	21 million	12.3 million	180,000

Tritech is serious about closing the loop, too. Tritech's order form even has a box to check if the clients want recycled stock for their printing jobs, which many do. The Site Manager estimates that his organization alone recycled more than 3,000 tons of paper in 1994. That translates into a lot of trees, energy, and clean air (see Table 6-3).

The Assistant Vice President of Facilities Purchasing negotiates the contracts for maintenance and services that support building operations. These include the haulers, recyclers, and vendors of janitorial products. He attributes the company's acceptance of janitorial paper products from recycled fibers to Andy Lauro and his team. Recycled toilet tissue and hand towels were tested in one building at a time. As Andy says, "If no one noticed, we went with it."

Is recycled paper more expensive? The Manager of Printing Services says that Merrill Lynch spends the same or slightly more money for recycled than for virgin stock. While prices are higher for individual purchases, she points out that buying by the truckload qualifies the company for volume pricing. True, Merrill Lynch has economies of scale on its side. But with increased demand, significant improvements in technology, and many more mills able to manufacture recycled paper, businesses of any size can make the switch affordably.

Isn't the quality of recycled paper less than that of virgin stock? She says that the paper industry has perfected techniques for remanufacturing uncoated paper. However, the ink holdout on recycled coated paper is not as good and has a flatter finish than virgin stock. It also tends to pick or flake. Merrill Lynch has found applications to use recycled paper without affecting the quality of the end product.

Does the end product look recycled? "Unless you've been trained, you can't tell the difference between virgin and recycled paper," says her Vice President.

On a Mission

Merrill Lynch recommends that its branches use paper and other products made from recycled materials, but doesn't mandate it. National Office Supply, its office products vendor, features recycled products in its catalog that it distributes to all Merrill Lynch departments.

At first some departments objected because they thought that it would cost more and that they would have to sacrifice the quality of the printed piece. They thought their literature wouldn't look its best on recycled stock. That's where people like the buyers in the Printing Services department came in. Despite objections, they promoted the use of recycled stock throughout the company.

One buyer has been a particularly successful advocate of closing the loop. She recalls that in 1990, Merrill Lynch bought hardly any recycled paper except for special printing projects. Volume picked up a little the following year. Then in 1992, she made it her personal goal to persuade the skeptics throughout the company to use recycled paper. Most of her internal clients didn't believe that the quality was as good as virgin stock. In meetings with department managers, she assured them that they wouldn't be compromising quality. She gained their confidence and convinced a few of them. This persuaded other clients to convert from virgin to recycled paper stock for their projects, too.

The other buyers also met with internal customers in groups and one-on-one. They assessed their needs and educated them about recycled paper. Then they worked with their suppliers, who researched the market to fit their needs (see Table 6-4). Their efforts paid off. By early 1995, up to 50% of the paper the firm used was from recycled stock.

Now in the most recent stage, Merrill Lynch has gone outside the corporate walls to educate other companies on the recycled paper market and closed loop recycling concepts.

Table 6-4.   Merrill Lynch's Sources for Quality Recycled Paper

Mill	Virgin Stock	Recycled Equivalent
S. D. Warren	Lustro	Lustro Recycled
Potlatch	Quintessence	Quintessence Remarque
Potlatch	Vintage	Vintage Remarque

Beyond Paper Recycling

Merrill Lynch also earns money for every toner cartridge returned for recycling through another vendor, NER Data Products.

The company even recycles its computer equipment under the auspices of the Assistant Vice President of Technology Investment Recovery/Facilities Purchasing, who started her career as a buyer. She is an active member of the Investment Recovery Association, a nonprofit international professional association. IRA members are Fortune 1000 companies committed to maximizing their companies' investments in capital assets and thereby helping to protect the environment. She considers economics to be the greatest obstacle in getting people to recycle. "Show businesses that it makes economic sense to recover and redeploy, and they will do it."

Early on, she noticed that some departments often retired the same types of computers that other departments were buying. Instead of processing the purchase requisitions, she redeployed the old computers into other parts of the company. At first she did this on a small scale, just a few at a time. Then around 1990, a single event changed all that and added a new dimension to the company's asset management.

Merrill Lynch was closing one of its facilities in Chicago, and she got the order to sell off all the equipment. She sent the word to the network of contacts she had developed that this equipment was available if anyone needed it. She got nine takers. Although it cost Merrill Lynch $17,000 to bring all of the equipment to the East Coast, this one transaction saved the company $125,000 in avoided purchase.

Today the firm uses college interns to recycle the company's retired technology. Merrill Lynch redeploys 25% back into the company with the ultimate goal of "zero landfill." It averages $1.3 million a year in cost avoidance with only three permanent full-time people and a few college interns. They deal with 6,000 to 10,000 serial numbers per year.

Merrill Lynch has built its reputation on its ability to make and manage investments wisely, and it practices what it preaches. In just a few years, its employees have made major strides in conserving resources. They have embraced Dan Tully's five principles in the way they do business and in the way they live. As a result, they have contributed to the company's and the Earth's bottom lines.

ENDNOTES

1.   *The Council on the Environment of New York City, Annual Report 1994* (New York, NY: CENYC), p. i.

2.   *Bullish On The Environment* (New York, NY: Merrill Lynch, March 1992), p. 1.

3.   "Recycling: cost-effective waste management," *Building Operating Management Magazine* (Milwaukee, WI: Trade Press Publishing Corp., October 1990), p. 32.

4.   This reflects a reduction in the amount of waste removed plus any and all recycling revenue received. The 1990 Cost Avoidance was based on the Department of Consumer Affairs maximum rate. A negotiated rate was used in developing the Cost Avoidance figures for 1991, 1992, and 1993.

5.   *The Earth Day Closed Loop Recycling Conference Program* (New York, NY: Earth Day New York, May 14, 1993).

6.   *Bullish On The Environment*, p. 2.

CHAPTER 7

GIORDANO PAPER RECYCLING CORPORATION

Humble Beginnings

Giordano Paper Recycling Corporation (GPRC) calls itself "The Paper People" and with good reason. Anthony Giordano, President, learned about paper recycling from his father and grandfather at A. Giordano & Sons in Newark, NJ. But he didn't take over the family business. His grandfather sold it upon retirement.

After trying to find satisfaction in various jobs, Anthony returned to the business he knew. With a $10,000 loan and one employee, he started his own paper recycling company in 1974.

His first big customer was a Fortune 500 company. Anthony saw how much mixed office paper it was throwing away. Waste disposal costs were climbing, and related legislation was popping up across the country. His competitors were collecting only high grade white and computer paper. Their labor costs were high because they hired people to manually remove all other paper-based items as well as the usual contaminants. Giordano wondered if mixed office paper could be put to better use and still turn a profit.

As if reading Anthony's mind, the Fort Howard Paper Company mill operator in Green Bay, WI, called him. With a new mill opening in Georgia, Fort Howard wanted a steady supply of secondary fiber. Anthony told the paper manufacturer that if it could handle sorted office paper, he could give them a dependable supply right from his own back yard. That began a synergistic relationship. Fort Howard developed processes to recycle office paper into other paper products, while Anthony prospected for customers who could supply the "raw" materials.

After about a year of trial and error, he finally succeeded in recycling mixed loads. He took all kinds of paper without his customers having to remove the rubber bands, staples, and paper clips. Anthony accepted even window envelopes, magazines, and junk mail. The paper mill mechanically sorted all the staples, paper clips, rubber bands, metal strapping, and baling wires after pulping. The metals salvaged in the process were also recycled, generating incremental revenue for the mill and saving landfill space. By 1985, Giordano and one mill were recycling over 300 tons of paper, newsprint, and corrugated per month. Over 100 tons of that was office paper.

Timing Is Everything

In 1984 while GPRC was still developing its processes, AT&T in Basking Ridge was looking for a solution to its waste paper dilemma. Giordano came to the rescue. Another synergistic relationship blossomed.

Anthony's goal was to provide a way of capturing the widest range of paper as conveniently as possible. He achieved this goal with his "two-can" recycling system: a large under-the-desk trash can for all paper, and a smaller garbage can to hang inside or outside the larger one. In the decade that Anthony has been propagating his two-can method, he has distributed more than 365,000 sets of trash cans to his customers.

He recalls the skepticism of eight of the ten AT&T task force members. They had been satisfied using his services for their office white paper recycling, and doubted that a mixed paper program would be better. How could this work and not cost AT&T money?

Anthony explained the economics. It would cost $50 per ton to send it to the dump. AT&T was getting $80 per ton to recycle white paper and saving on disposal costs. Mixed paper was paying only $20 per ton. Would the corporation lose $60 per ton by commingling? Anthony showed task force members that by capturing 90% of what they were throwing away, they'd make up the difference in volume and then some.

It didn't take long for him to produce results. He estimates that he gets an average of 2–3 pounds of recyclable paper per person per day.

Anthony's prospective customers are also often skeptical that there's any money in recycling mixed paper. He refers them to 30 or 40 of his largest customers. Just a few phone calls and testimonials of the high recovery and reimbursement rates win them over.

GPRC has become the unequivocal recycler of choice among Fortune 500 companies in New Jersey, Pennsylvania, and New York.[1] In 1995, New Jersey Governor Mrs. Christine Todd Whitman recognized GPRC as one of the top-twenty privately held corporations in northern New Jersey with which to do business. The company also received the Award for Outstanding Achievement in recycling from the New Jersey Department of Environmental Protection for five consecutive years.

Sometimes competitors try to woo away his customers and prospects by offering the second trash can free. Anthony points out to them that nothing is free. Either the cost of the cans will be built into their monthly payments, or the suppliers will remove the cans when their contracts expire even though the customer has already paid for them. Anthony's customers recuperate the low start-up costs within the first few weeks of using the two-can system.

Today, AT&T and other GPRC customers recycle more than 90% of their waste paper by following this simple system. With over 60 employees and growing, the distinguished recycler boasts a 3,000–4,000% increase in volume over 1985 when he started his office paper recycling program. Now he recycles paper, newsprint, and corrugated in excess of 10,000 tons per month, of which 4,000 tons are sorted office paper.

Anthony gets very clean loads of paper from his customers. With the two-can system, food containers and scraps are not usually a problem, although contaminants such as candy wrappers and waxed boxes do creep into the mix from time to time. GPRC workers manually sort this waste while the mills extract the rubber bands and metals. Of the average 10,000 tons collected each month, only about 0.8% or 80 tons are actually thrown away.

Anthony offers a confidential document destruction service, either at the customer's site with a portable shredder or at GPRC. "Giordano

Paper Recycling Corporation has installed a new paper shredder capable of destroying up to 5 tons per hour…in a secure area monitored 24 hours a day by guards and close-circuit TV….Customers can now witness the destruction of their materials in the comfort of an enclosed office in full view of the entire process."[2]

GPRC's recycling specialists stay current on recycling and waste legislation. They evaluate the customer's site, suggest the most efficient, economical recycling program, set up internal procedures, and provide educational materials.

He Who Laughs Last

When Anthony first started his mixed paper recycling campaign, his competitors laughed at him. Now Anthony and his customers are the ones who are laughing with the satisfaction of knowing how many resources they've saved and the reimbursements and cost avoidance that have gone straight to their bottom lines. Being a pioneer and getting into the market early has put Anthony way ahead of the pack with a solid customer base and a network of partnering vendors.

Crystal Ball

What does the future hold for recycling? Anthony sees a bright future with increasing volumes and continued growth.

He says that all the major mills have converted to newer technology that allows them to economically recycle all grades of paper. Others will have to do the same within the next few years to be competitive, comply with federal mandates, and stay in business. Each year the percentage of mandated recycled fiber should increase.

He sees smaller companies, multi-tenant buildings, doctors' and dentists' offices, clinics, and hospitals as another source of "raw" materials. Pooled recycling allows multi-tenant building managers to rent space cheaper because they're being reimbursed and avoiding maintenance costs.

Anthony usually sees a slowdown and lower reimbursement rates in the summer as U.S. mills close for vacations and clean-ups. The entire

European market also slows for August vacations. But this is only temporary, and the market generally picks up again in September.

More importantly, he anticipates growing global demand for paper and recycled fiber. He says that most virgin fiber for paper comes from tree farms and that with today's progressive agriculture, it takes only 15 years to grow a tree instead of 75. "But why use trees to make paper if we can make paper from paper?" asks Anthony. "Use the tree farms to supply materials for other wood products like furniture and lumber, and save our forests and old growth trees."

"Over the next year, Americans will throw away 28 billion glass bottles, 18 billion disposable diapers, 1.8 billion pens, 247 million tires, 2 billion razors and blades, 12 billion mail order catalogs, and enough aluminum to build 30 airplanes. This is why recycling just makes sense. And you'll feel good about it, too."[3]

As for closing the loop, "shopping in many ways is just like voting. Each purchase you make sends a message to the manufacturer that you 'vote' for their product. As a result, the product manufacturer continues to make the product in the same way.... By purchasing products made from recycled material, you support recycling programs around the country."[4]

GPRC also does business in New York , Virginia, and North Carolina. Giordano plans to expand into Chicago, San Francisco, and Los Angeles—proof that one person's trash is another one's treasure.

Why does Anthony Giordano believe any company, large or small, should implement a paper recycling program? "Because it's profitable for the customer and it's the law."

ENDNOTES

1. According to Giordano Paper Recycling Company.
2. Giordano Paper Recycling Corporation's on-hold message (Philadelphia: The Hold Company, Inc., July 1995).
3. Ibid. However, not all old growth forests are replanted with the same species.
4. Ibid.

CHAPTER 8

V. PONTE & SONS

A Century of Service

Unity Environmental helps its New York City clients achieve their waste reduction goals. Its affiliate company, V. Ponte and Sons, Inc., has been in the recycling business since it was founded in 1919, paying local businesses and office buildings for their waste paper. Through vision, hard work, and self-motivation, the four Ponte brothers grew and diversified the company. One of the largest waste haulers in the Northeast, the firm owns a paper recycling facility. Its paper mill in Clifton, NJ, manufactures paperboard from recycled stock. This paperboard takes the form of tubing, partition stock, and boards and game boxes for Milton Bradley, and cereal, shoe, and gift boxes for other major companies.

Partnering for Success

Unity sells its waste paper hauls to competitors' mills as well as its own paper mill, usually on an individual basis. It picks up its clients' waste paper and corrugated cardboard and delivers it straight to the mills. What it might lose on reduced hauling volume it can make up on the lower costs of raw materials for its mills.

Clearing the Hurdles

The obstacle the Pontes have encountered most often is that people are creatures of habit. They don't like to change and give up what is familiar.

The most effective ways to get people to change their habits are to educate and retrain them. This is the Ponte family's biggest challenge. For example, separating high grade paper like white ledger and greenbar from mixed paper brings in higher revenue. But that requires consistent cooperation. Sometimes the wrong things like newsprint, groundwood, polystyrene, plastics, and food end up in the wrong bins. People get busy or distracted, make mistakes, or sometimes just don't care. Most companies ignore recycling when the effort detracts from employees' jobs.

What's Ponte's solution? Awareness and simplicity. These are the key ingredients to a successful waste reduction and recycling program. Clients handle the awareness. As for simplicity, Ponte recommends co-mingling. While others may recommend two cans, Ponte prefers the one-can system as the only viable long-term solution: a single can for waste paper—a true "waste paper basket."

V. Ponte & Sons advises its clients not to use plastic liners in the employees' waste baskets. Eliminating the plastic liner discourages people from throwing anything but recyclables in the bin. Employees must get up and throw out real garbage in a centrally located container. This procedure accomplishes several goals:

- Employees take more breaks, diminishing the risk of repetitive task syndromes and job-related stress.

- The company avoids the cost of plastic liners and additional trash containers for each employee.

- Food waste and odors are more easily controlled, minimizing the risk of insects and other pests and making the work environment more pleasant.

Even the residential market is gravitating toward collecting and recycling 100% of all waste paper, not just newsprint. "Easiest is best," says Ponte. That's why the company allows and even encourages co-mingling

of all recyclables—metal, glass, paper, and plastic. The MRF sorts out
the garbage and separates the various grades of paper. The paper then
goes through a trammel to take out the grit and dirt. What's left is made
into everything from tissue to desktop paper. Polystyrene, however,
poses a handling problem and is beyond the scope of this book.

Ponte goes on to say, "The market has exploded because of concern
about whether there'll be enough paper for the recycling mills that are
coming on-line." This gives waste generators a greater profit motive to
recycle more of their waste paper and cardboard.

Unity Environmental has the program so refined that almost over-
night they can set up a program for a corporate client to recycle 85% of
all waste generated. For every 40 yards of a typical client's office waste,
only about 4 yards or 10% will go to the landfill.

Up in Smoke

What about incineration, the route that Florida has taken? The
Pontes call it "economic suicide. Florida has built huge incinerators—
waste-to-energy plants—and forgot about recycling." They say that the
main concern has become keeping a steady supply of fuel for the
incinerators, supporting consumption, and defeating the basic purpose
of reducing waste and conserving resources. They believe that the
incineration companies are concerned with the public's "not in my
backyard" attitude to the toxic fumes inevitably emitted. In contrast,
paper mills emit fewer toxins, use less energy and water, and are much
easier on the overall environment when using recycled fiber as raw
material.

A Long-Term Investment

The Pontes recall that in the early years of Earth Day, some companies put a lot of energy and money into recycling, and lost a lot, too.
In 1992, waste paper sold for less than it did in 1942, not adjusting for
inflation. By early 1995, prices were way up due to a healthy waste
paper market and mills equipped to handle it. "The secret is in marketing and creating uses," say the Pontes.

Why recycle? The Ponte family says, "Simple economics."

PART III:

HOW DO WE GET OUT OF THIS MESS?

CHAPTER 9

THE ANTIDOTE

The Challenge

How do we counteract our destructive influence on the environment? Stabilizing and even reducing the rate of population growth are long term vital strategies. In the short run, we can reduce consumption and be more selective with our choices "to print or not to print."

The Three "Rs"

"Source reduction tops the list of solid waste management options adopted by the U.S. Environmental Protection Agency and endorsed by most states and localities."[1] Reduce by consuming less in the first place. EPA's other options in descending order of priority are reuse, recycling of materials, waste combustion, and landfilling. Reuse goods as much as possible, or let other reuse items that are still functional. The third "R", recycling what we discard, is our last opportunity to keep valuable resources alive. That's why buying products made of post-industrial materials and closing the loop is so important.

Some people believe that it takes legislation to achieve waste reduction goals. "Waste reduction as we see it means packaging restrictions, limits on the development of new disposable products, as well as reductions in toxic constituents in products that ultimately find their way into the municipal waste stream."[2] Others, like McDonald's, suggest avoiding legislation by addressing "the issue of excess packaging, packaging uses, and packaging recyclability or content."[3] Still others emphasize the importance of designing products for recyclability.

The Fourth "R": Redesign

The key to all of our solid waste management options is cradle-to-grave product design.

> "Clearly, if industry is to provide for today's and tomorrow's needs without undue environmental degradation—the goal of sustainable development—new processes will be needed that use less virgin material, produce markedly less pollution or waste per unit of product, and minimize indirect environmental effects....Worldwide, more than 1,100 firms have endorsed the Business Charter for Sustainable Development, setting forth broad environmental principles for corporate conduct."[4]

Paper manufacturing technology has embraced this concept. It has made big strides in improving the bleaching process and producing ECF[5] and TCF[6] paper products. Mills emit far fewer dioxins (associated with chlorine) than just a few years ago.

> "Design for recyclability" suggests that inks and glues should be made with recycling in mind. "Synthetic glues that are commonly used in binding books and magazines can create grave difficulties by causing a sticky residue in the paper-making plant. Some modern inks are actually formulated to resist dispersion in water. Designing new products for easier recycling at the end of their first life can help resolve these issues, and maximise reuse opportunities."[7]

Meeting the Challenge

Redesign and source reduction are imperative in alleviating the garbage dilemma. Reusing the end products until they have outlived their usefulness means getting the most out of our resources, both renewable and nonrenewable. Once the waste is generated, recycling (which includes buying recycled goods) is our last chance to keep valuable materials out of the waste stream. Chapter 10 addresses and enumerates the organizational strategies and actual tactics of implementing a successful paper recycling program.

ENDNOTES

1. *The Solid Waste Dilemma: An Agenda for Action* (Washington, DC: U.S. EPA, February 1989), p. 16–19 (cited in Bette K. Fishbein and Caroline Gelb, *Making Less Garbage: A Planning Guide for Communities.* New York, NY: INFORM, Inc. 1992), p. 2.

2. Sept. 21, 1991, Senate hearing on the Waste Management Provisions of S.976, the Resource Conservation and Recovery Act Amendments of 1991 before the Subcommittee on Environmental Protection, Committee on Environment and Public Works; Randall Franke, County Commissioner, Marion County Oregon; National Association of Counties; National League of Cities, Solid Waste Association of North America.

3. Steven M. Polan, Commissioner, New York City Department of Sanitation. April 24, 1991, House hearing on state and local recycling programs before the Subcommittee on Transportation and Hazardous Materials, Committee on Energy and Commerce.

4. *World Resources 1994–95* (New York, NY: Basic Books), p. 214.

5. Elemental chlorine free.

6. Totally chlorine free.

7. *The Warmer Bulletin, Number 43, Information Sheet: Paper Making & Recycling* (Tonbridge, Kent, UK: The World Resource Foundation, November 1994), p. 4.

CHAPTER 10

CORPORATE ACTION PLAN

We've diagnosed the problem and seen the many solutions and sizable benefits that some leading companies have employed. Now, how do we sell the idea to top management as well as to the employees? And what specific strategies can we employ once we've sold the idea?

In their book *Organizational Unconscious: How to Create the Corporate Culture You Want and Need*,[1] Robert F. Allen and Charlotte Kraft suggest a four-phase approach to changing or developing a company's culture, belief system, or spirit. They begin with research and planning. This entails analysis and setting objectives. The next step is to involve people by system introduction and involvement. The third phase is system implementation and change. The final phase is evaluation, review, and extension.

From the principles presented in *Management*,[2] one can also develop a four-phase approach: Planning and decision making, organizing for stability and change, leading, and controlling. *Management: A Book of Readings*[3] advocates a similar "managerial transformation process" whose main steps are planning, organizing, staffing, leading, and controlling.

We can apply the basic concepts from these three books to the goal of minimizing waste and maximizing resources and profits. The following plan is more applicable to large organizations than to smaller ones, although many of the strategies and tactics are useful regardless of size.

Phase 1: Research and Planning—Where Are You? Where Are You Going?

This initial phase involves three main steps. First, investigate the situation. Next, analyze what you've learned. Then, determine your goals.

Investigate

Determine your trash volume (see Figure 10-1). If your disposal costs are factored into your rent as is done in industrial parks and office centers, ask the property management company for the information. Your waste hauler can give you a breakdown if you have one, or estimate how much is being generated using the following formulas:

Tons of Compacted Waste = Size of trash container (cubic yards) x 400 pounds per cubic yard ÷ 2,000 pounds per ton

Tons of Uncompacted Waste = Size of trash container (cubic yards) x 100 pounds per cubic yard ÷ 2,000 pounds per ton

Containers	Number of Containers ^{x}	Capacity (cu yds) ^{x}	Frequency of Hauls ^{x}	% Filled when Hauled $^=$	Monthly Volume
Compactors[4]					
Dumpsters					
Other					
Total					

Figure 10-1.   Trash volume worksheet.[5]

Calculate current waste disposal costs. This may be the most valuable information you'll gather to determine the return on investment (see Figure 10-2), because there is more to be gained by reducing waste volume and related disposal costs than by being paid for recyclables.

Containers	Capacity (cu yds) ^{x}	Average # of Hauls per Month ^{x}	Cost per Haul[6] $^=$	Cost per Month
Compactors[7]				
Dumpsters				
Other				
Total				

Figure 10-2.   Disposal cost worksheet.

Conduct a waste audit. Find out about local recycling regulations so that you'll know the opportunities you need to identify as the waste audit progresses. There are two ways to conduct a waste audit.

One way is to do a visual inspection of inside containers and outside dumpsters. This is certainly the easier of the two ways and will provide most companies with a pretty good idea of their waste streams.

The other way is to pick through both inside containers and outside dumpsters, i.e., sort through the trash. If this sounds too distasteful, call in an expert. There are consultants, government agencies, and waste haulers who have staff to do this. While this may be more accurate than the first method, it is likely to produce similar findings and be more trouble than it's worth. Also, if you need management's approval to do a detailed a waste audit, you may prefer the first technique.

If you do decide to do a pick-through, get help. Have everyone wear protective clothing—rubber gloves, goggles, aprons, and boots. You'll need to put the separated trash into plastic bags or containers, so have plenty on hand. Sort the trash by recyclables, cafeteria waste, non-recyclables, and reusable items. Non-recyclables include paper products meant for disposal such as napkins, paper towels, and tissues, as well as paper that is wet or contaminated with food waste. Do this for each type of facility since the waste profile for a manufacturing plant will differ greatly from that of an office environment (see Figure 10-3).

Determine what recyclable materials are worth recycling. Sometimes transportation methods, costs, and other factors make recycling more costly to the environment than disposal and need to be taken into account. Consider environmental laws and penalties. Also, research the demand for your recyclables (see Figure 10-4). The overseas market is extremely important. Timber-poor countries like Japan are eager customers for recycled paper. If the markets are soft, make the commitment to become a consumer of the end product. Products with post-consumer content may cost a little more than those made of virgin material until more businesses and consumers demand recycled content. As demand grows, prices will drop until eventually they even out. Paying a slight premium up front will pay off for everyone in the end.

Once you determine what recyclable materials are marketable, look for a customer. If you don't generate a large volume, you may be able

	Daily Weight ÷	Pounds per Month =	Percent of Total
Recyclable Material			
Magazines, newspapers			
Office papers (computer paper, fax papers, junk mail, mixed paper, printing and writing papers, sticky notes)			
Packaging			
Pallets			
Other (specify)			
Subtotal			
Recyclable Cafeteria Waste			
Aluminum cans			
Bi-metal cans			
Glass bottles and jars			
Packaging			
Plastics (bottles, containers, bags, wrap)			
Polystyrene plates, cups, etc.			
Food surplus (still fit for humans)			
Subtotal			
Non-Recyclable Waste			
General cafeteria trash			
Disposable and contaminated papers			
Subtotal			
Reusable Items (itemize separately)			
Subtotal			
Grand Total			

Figure 10-3.   Waste audit worksheet.

to team up with other small companies. Businesses in multi-tenant buildings can generate enough volume together to be attractive clients for recyclers. Larger companies in the area may let you piggy-back on their programs, too, as AT&T did in Atlanta. If your volume is great enough, you may be able to forge an agreement with the mill or manufacturer.

Many companies, large and small alike, prefer to deal with a recycler, broker, or consultant who can provide a total solution. Ask this business partner about source separation and storage needs. Special conditions may include preparing the materials for pick-up or shipment, storage considerations, or exceptions to terms of the agreement. Find out who will manage the pick-ups and compile the reports. Who will supply the containers? Will you be charged for them?

Material[8]	Annual Volume[9]	Price Range	Hauling Cost	Special Conditions

Figure 10-4. Market development worksheet.

Reimbursement amounts are very volatile. Don't be surprised if you get $200 per ton of corrugated one month and only $50 the next. Remember that you have long-term goals and that cost avoidance is key.

Table 10-1 shows estimated market prices for certain paper grades in the Chicago area over the last ten years.[10] These are meant as a guideline and in no way reflect published or negotiated rates. Parentheses indicate that waste generators had to pay to have the goods hauled away, possibly due to a temporary glut in the market. Although this may happen from time to time, it still costs less than most waste disposal fees.

So how do you know whether or not you're getting a fair price for your waste paper? Many companies subscribe to various publications and newsletters like *The Fibre Market News* and *Pulp & Paper Week* to get an idea of average prices around the country. Be aware that the rates are estimates to be used only as a guide and may not accurately reflect what your organization can earn.

Table 10-1. Approximate Waste Paper Reimbursement Rates in Chicago, IL[11] (dollars per ton)

	July 1985	July 1986	July 1987	July 1988	July 1989	July 1990	July 1991	July 1992	July 1993	July 1994	March 1995	June 1995	July 1995
White Ledger	40	N/A	40-45	90	80	50-55	50	60	40	45-50	195	200	130
Colored	30	N/A	25	80	70	40	30	20-25	30	25	70	70-90	70
CPO	N/A	N/A	60	110	115	65-70	70	80	85	95	200	220	210
Mixed	2.50	N/A	2.50	5	0	(10)	(20)	(35)	(10)	10-15	15	15	15
Corrugated	25	N/A	40-45	20-25	0	10	5	(5)	15	40-45	100	80	80

	Current Disposal Cost[12]	Cost with Recycling	– Income from Materials	= Net Savings by Recycling
Container Cost				
Hauling Cost				
Contract Cost				
Total				

Figure 10-5.   Monthly cost savings worksheet.

Figure out the bottom line. This is the moment of truth. Is there money to be made by selling recycled paper and corrugated? Will the bottom line be affected more by lower consumption and resulting savings? What waste disposal costs will be reduced? Cost avoidance alone may be able to justify the project (see Figure 10-5).

This isn't the end of your research. You'll need to look at how the organization works before setting objectives and structuring the plan.

- **Examine current processes.** Pinpoint sources of internally generated waste. Simply re-engineering how things are done can make a big difference in the waste volume. Chapters 4, 5, and 6 contain examples of re-engineering, such as making two-sided copies, using electronic mail, and training employees to use on-line reporting systems.

- **Identify available or additional technology and other resources to reduce waste.** Does the company already own a baler or compactor? Does it make sense to purchase or lease one? What else is needed?

- **Reduce and reuse.** Look for opportunities to conserve.

- **Close the recycling loop.** Substitute recycled products for those made with virgin fiber. Scrutinize internal policies and purchasing guidelines to see where recycled materials and reusable goods may be used instead of new ones.

Set Realistic Objectives

To measure success, you must have a baseline (see Figure 10-6). When setting both short- and long-term goals, be sure to address financial aspects beyond cost avoidance, such as additional equipment and personnel costs. Consider, too, company image and government regulations.

Make objectives as specific as possible to avoid misinterpretation and to facilitate measuring levels of achievement. Assign time lines to keep the project on track.

Think about participating in the National Office Paper Recycling Challenge. It offers a specific, easy-to-follow, effective methodology. Your company may even distinguish itself by winning an annual recycling award.

Keep an open mind and a sense of innovation. Here are some sample goals and objectives:

- Recycle __% of waste generated using ____ as the baseline year.
- Reduce hauling fees by __% by ____ (month and year).
- Purchase disposable paper goods with at least _% post-consumer content by ___ (month/year).
- Purchase printing and writing papers with at least _% recycled fiber by ___(month/year).
- Decrease paper consumption by __% by ___ (month/year).

Draw up a formal proposal and update it regularly . If your organization is large enough, a recycling committee can plan and execute the program. Otherwise, management and the recycler can handle it.

Action Areas	Priority (A, B, C)	Person(s) Assigned	Goals	Actual

Figure 10-6. Waste reduction goal worksheet

Analyze Crucial Areas of Influence

An early and visible win will increase your chances of long-term success. Start in an area where the results can be measured and seen immediately by a lot of people, particularly by management. The most impact may come from curbing consumption by duplexing lengthy documents, or from reducing disposal costs by recovering and recycling waste paper.

In the community, enlist those who have similar interests. Establish an informal network of people who have already worked on recycling and waste minimization programs, as well as those who want to get involved. The former will have valuable experience that will save time, effort, and money. The latter will have energy and enthusiasm. Check with your local department of sanitation or chamber of commerce.

Network with others in your industry as well. Together you can develop strategies that address common issues. Take the lead and become a role model for your customers, suppliers, and competitors. The positive publicity will have a positive effect on your bottom line and market share, as McDonald's has proven with McRecycle USA.

Develop Task Forces and Committees

It is critical that those who are responsible and accountable for developing and executing the strategies have decision-making authority.

By definition, a task force is a group of people organized to achieve specific objectives and then disbanded. The task force will do the research, recommend distinct strategies, and get the program going. Make the janitorial service an integral part of the task force. How well the cleaning crews execute their duties is one of the most important determinants of success, in addition to top management support and employee awareness.

While a committee is similar to a task force, it is more enduring and designed to pursue long-term goals. The committee will keep things going in a large organization. In a smaller one, a recycling coordinator, consultant, or manager may be enough.

Designate key individuals with the responsibility, as well as the authority and accountability, for making things happen. These may be just those who serve on the initial task force or may also include recycling coordinators or committee members. In either case, group members should come from a variety of areas within the organization. By touching all departments, good habits will continually be reinforced. Any employee with a question about the program should know who to call for quick and easy access to the answers.

Long-term recycling coordinators who are respected individuals "within the ranks," like Cheryl LaPerna at AT&T in New Jersey and Marilyn May at AT&T Power Systems, will be able to enroll a lot of people quickly. If top management isn't fully behind the program, these individuals can propel a grass roots movement. Ultimately, though, support from executives is essential.

Managers distinguished for their people skills, like Andy Lauro at Merril Lynch and Jerry Twardy at AT&T in New Jersey, can contribute in several areas. They can help everyone accept and institute change more readily and keep the program visible. They will provide a balance for managers who are not interested, don't support, or sabotage the program. They will help promote the goals and achievements to upper management. This is particularly important if the program does not have the full support from executives.

Volunteers with emotional commitment, like Richard Vickers, Ron McCauley, and Phil Lombard at AT&T Power Systems, and George Perry at AT&T GRE, will sustain the program when interest and energy wane. Choose individuals who believe in the basic philosophy and are willing to take responsibility for instituting change. Include believers from management, like McDonald's Shelby Yastrow, Bob Langert, and Mark Brownstein.

Recycling companies and consultants can be invaluable members of the task force and on-going committees. They have the experience in what works and what doesn't, the statistics to validate return on investment, the contacts to get the best rates and services, and the resources to deploy effective systems and processes. They're available to increase employee awareness and stimulate team spirit.

How long the volunteers serve on a task force or committee may vary from one organization to another. Marilyn May believes that three month "terms" do not give committee members enough time to accomplish many of their goals. For AT&T Power Systems, allowing volunteers to serve on committees as long as they want works well. When one of them leaves, he/she nominates a successor. An alternative is to replace only a few committee members after each term in order to provide continuity and leadership.

Solidify Commitment

Keep the Generals on your side and the troops rallied behind you with firm commitment:

- **By employees.** Find out what's important to employees. If the company has a profit sharing or bonus plan, some employees will be motivated by the prospect of higher dividends. Arrange for a portion of the income to be donated to favorite charities. To lock in commitment and to thank everyone for their participation, an occasional party or employee event may work.

- **By department managers.** Set waste reduction, consumption, and recycling goals for all levels of management as part of their performance review. While most managers will want to do the right thing, the cash incentive will work for the few who may not care.

- **By top management.** Even the best grass roots campaign will falter without the commitment of upper management. As we've seen in every case, adopting the four Rs—reduce, reuse, recycle, redesign—can add up to substantial incremental income, lower costs, and positive press and market share. These key benefits will get the executives' attention. It helps if some of them also believe in the cause.

- **By the janitorial service.** Properly inform, train, and monitor the cleaning crew so that your efforts don't literally go to waste.

- **By vendors.** Recyclers and consultants have the monetary incentive to make sure the program succeeds. Existing suppliers can also reap benefits. While McRecycle USA worked wonders for McDonald's and its vendors, you can start small. Ask suppliers to cooperate by reducing the amount of packaging, increasing the amount of recycled content in their products and packaging, and using reusables. You'll be surprised how many of them will agree on a volunteer basis. If necessary, rewrite company specifications to nudge reluctant vendors in the right direction.

- **By community leaders.** Get involved with local community groups like your local Chamber of Commerce. And get them involved in your planning process. What's in it for them is the community's enhanced image as an attractive place to live and do business.

Formalize the Plan

The plan is most effective when it is written, has quantifiable objectives and time lines, is approved and supported by top management, and is communicated regularly throughout the organization. As stated earlier, it may include performance measurements and rewards for managers, departments, and employees.

Clearly laid out strategies and tactics will leave little to chance. The case studies in this book have provided a wealth of examples. Many such strategies, although not all, are summarized in Phase 2.

Phase 2: System Introduction and Involvement—How Are You Going to Get There?

Gather all your resources and line up the troops. The road to success has its pitfalls. Anticipate them and prepare alternate strategies.

Strategies and Tactics

Reduce

- **Review office paper practices.** "Most desk-oriented employees generate approximately half a pound of paper waste a day."[13] This may be higher in computerized organizations. Giordano Paper Recycling Corporation and AT&T

in New Jersey estimate that each office worker can generate as much as 1.5 to 2 pounds per day.

- **Print on both sides of the paper.** Use paper already printed on one side for drafts or as notepaper.

- **Write short notes and memos on small pieces of paper or on narrow-lined, two-sided notepads.**

- **Consider upgrading or replacing laser printers and copiers.** Change them so that they print over sixty percent of their copying double-sided, are more energy efficient, and don't emit harmful ozone. Train employees to use the duplexing feature. "Columbia University has made double-sided copying the default mode on their copiers. Over 60 percent of their copying is now double-sided."[14]

- **Reduce computer printouts and reports.** Prepare documents double-sided and single-spaced. Adjust margins to avoid pages with little text.

- **Store files electronically.**

- **Start or increase the use of electronic mail.** Train and encourage computer users to read electronic mail and reports, edit drafts, and proofread documents on-screen, printing only the pertinent pages when necessary.

- **Survey report recipients.** Identify those who do not need or no longer wish to receive various reports, or need only a part or a summary of certain ones.

- **Develop an on-line reporting system.** AT&T Power Systems in Mesquite, TX, decreased the number of printed reports 73% from 1986 to 1995 by asking employees to use the on-line reporting system. "Kinney Corporation queried its computer report recipients about whether a summary would suffice...(This) dramatically reduced the volume of reports printed, saving Kinney over $200,000 annually."[15]

- **Eliminate header pages and slip sheets.** Blank sheets of paper separating print jobs are a big waste of money. Reprogram equipment to eliminate header pages and slip sheets. If some users require these, post signs reminding them to put the blank pages back in the printer tray for reuse.

  Columbia University saved 75,000 sheets of paper, about a dozen toner cartridges, and over $2,000 a year in paper and toner costs by changing the default setting on 18 of its 45 mainframe printers to eliminate header sheets. Kinney eliminated blank pages that were automatically placed between print jobs, realizing a savings of over $30,000 annually.[16]

- **Cut down on in-house mail.** For internal publications, reports, and memos, periodically attach a note asking those who do want to keep receiving them to notify the originator by a specified date. Remove the names of those on the distribution list who do <u>not</u> respond. Human Resources or Payroll can update distribution lists since they are among the first to be notified when employees are hired or terminated.

- **Broadcast short messages via voice mail.**

- **Establish a central file to reduce the number of individual copies generated.** Use it for internal documents and reports, publications, reference material, and telephone directories. Have each department circulate a single copy or just a few copies for department members to read, initial, and return to the manager or secretary.

- **Focus on reducing the number of publications that employees receive.** Eliminate those that are unnecessary. Cancel multiple subscriptions where one or a few copies will satisfy the needs of individual departments or work groups. Encourage employees to cancel unused subscriptions. You'll save many hours of mailroom personnel costs as well as trees.

- **Trim written communications to employees.** Use bulletin boards conveniently located near employee entrances and common areas. Make the most of electronic mail and electronic bulletin boards for those employees who use computers.

- **Utilize reusable envelopes for interdepartmental mail.**

- **Does your company celebrate Christmas with a company tree?** Have it mulched after the holiday. Better yet, buy a live tree and donate it to a nearby forest or park for replanting after the holiday.

- **Send faxes from a personal computer.** You'll reduce paper and postage costs. If that same fax is sent to another personal computer, there's no paper generated at all. Even better, correspond via the information superhighway using the Internet, CompuServe, America Online, Prodigy, and other electronic services. You'll communicate faster, save money on envelopes, paper, and postage, and avoid tedious time-consuming filing.

- **Order only writing and copier papers that are processed without chlorine.** This will reduce toxins.

- **Develop a plan to replace environmentally harmful products and practices with environmentally preferable alternatives.** For example, have the cleaning crew use sponges and mops instead of paper towels (and benign cleaning agents, too). Instead of chlorine-bleached coffee filters, use filters that are 100% oxygen whitened, unbleached, or reusable. The same goes for toilet tissue, paper towels, and napkins.

- **Reduce the amount of toilet tissue and paper towel waste.** "Use bulk paper towel dispensers which regulate quantity used instead of loose roll paper towels."[17] Large rolls of toilet tissue will also help reduce waste as well as packaging.

- **Reduce the volume of packaging materials.** Manufacturers can follow McDonald's lead and work with suppliers to reduce the amount of packaging material they use to deliver goods to you. In turn, reduce the quantity of your own packaging for your customers. Eliminate unnecessary padding and balance the durability of the packaging with the load it will carry.

- **Retailers can give customers incentives to carry items in reusable bags.** Some supermarkets in California give a $.05 credit for every box or bag the customer supplies, whether canvas, paper, or plastic.

- **Reduce the use of disposables.** Company cafeterias can investigate whether or not circumstances (patrons, costs, infrastructure) make reusable washable utensils better than disposables. The findings may suggest giving incentives such as a slightly lower price to those who use washable plates and utensils and bring their own beverage containers. Installing dishwashing equipment and buying reusable plates and utensils can pay for themselves in lower waste hauling fees and fewer supply costs.

- **Junk the "junk mail."** The California Department of Conservation, Division of Recycling, estimates that American homes receive about 4 million tons (English measure) of junk mail every year. The Warmer Bulletin's February 1995 issue puts that figure at almost 2 million tons (metric). About 44% of that is never opened or read. The junk mail Americans receive in one day would produce enough energy to heat 250,000 homes annually, and that doesn't include the 4.3 billion pieces of junk mail that fail to reach the addressees each year. Because mailing lists are sold for direct marketing to key decision makers, junk mail has penetrated the workplace, too.

- **One way to reduce junk mail is to avoid being on various mailing lists.** To do this, you can notify the Direct Marketing Association in writing to remove yourself from its mailing list. However, be sure to include as many iterations of your name, title, and address as possible. Be wary of sweepstakes, contests, and special offers. They're usually marketing ploys to get you on—you guessed it—their mailing lists. Most credit card companies periodically survey their customers for their preferences. The survey forms usually allow you to designate that you do not want to be on mailing lists that are sold to other marketing firms. When you do get junk mail, you can make your request by letter, phone, or fax; or simply write on the unopened material "Return to Sender" and let your post office do the rest. Eventually, when they've wasted enough postage on you, they'll get the message.

- **Maximize mailings.** "Mailing practices are notorious waste generators. Take a typical mailing to 1,500 clients: the print run ordered will most likely include an extra 100–200 copies, if not 500(!), just in case some additional copies are needed. Specify a modest number of back-ups."[18] Review the history of your mailings to see just how many extra copies you need.

- **Insist on two-sided printing.** "If the mailing is two or more pages and not printed on both sides, reams (perhaps cartons) of additional paper are used. Make it company practice to use both sides for printing and copying."[19]

- **Update mailing lists.** Remind departments that do mailings (sales, marketing, advertising, public relations) to be selective in whom they target and to audit and purge mailing lists regularly. "If the labels have not been recently updated, the (mailing) list may include names that should no longer be on it as well as incorrect addresses."[20] Correct or delete addresses on mail that has been returned as undeliverable. "Use Direct Marketing Association's Mail Preference Service computer tape to identify and eliminate consumers (prospects) who...do not wish to receive (information on) promotions. Use zip (code) correction programs."[21]

- **Stuff the envelope.** "If different departments send mailings to the same clients, announce mailings so others (in the company) may piggy-back onto them."[22] This will also save postage and costs for additional envelopes and labels.

- **Get customers to report duplicate mailings.** "Two large mail-order firms, Fingerhut and Heart Song, offer gift certificates as customer incentives for reporting duplicate mailings."[23]

- **Use two-way envelopes.** "If you send out return reply envelopes frequently, two-way envelopes may interest you. The same piece you send to your clients can be adjusted slightly and sent back to you—an extra flap is printed with the return address inside the envelope; it folds over the front to make a fresh envelope. The Illinois Secretary of State is reported to have saved over $50,000 in mailing approximately two million return envelopes. The printer (who) bid for the two-way mailer was $57,547 lower than the lowest bid for a two-envelope system. Central Hudson Gas & Electric in upstate New York has been using two-way envelopes for years, keeping two million unnecessary envelopes out of the waste stream. Fred Roguish, from the New York Times, a user of two-way envelopes, cautions, however: 'If your customer opens the mail with a letter opener, they've killed it. Also, the envelope is not readily manufactured, so you are limited as to where you can purchase it; if (you) need an order in a hurry, you will have difficulty'."[24]

- **Replenish the source by "re-tree-bution.** Another suggestion for businesses that deal in large mailings is to have trees planted to replace what you are using."[25] Use the following formula revised from one developed by Seventh Generation, a direct mail business, to calculate the number of trees used per catalog, report, or mailing:

First, find out the weight of paper used (WPU) from the invoice.

Next, multiply the WPU by the percentage of virgin fiber (deduct the pre-consumer and post-consumer recycled content).

Then divide the weight of virgin fiber by 2,000 to determine what percentage of a ton of virgin fiber was used.

Finally, multiply the result by 17 trees.

Example:

Weight of Paper Used (WPU)		500	pounds
Percent of Virgin Fiber (pre- and post-consumer @ 45%)	x	55%	
Weight of Virgin Fiber		275	pounds
One Ton	÷	2,000	pounds
		0.14	
Trees per Ton of Paper	x	17	
Number of Trees to Plant		2.38	trees

- **Reduce packaging.** "Packaging and containers make up approximately one-third of New York State's landfill contents."[26]

- **Economize on outgoing packaging.** Use a heavy-duty recycled paper cover for bound publications instead of a plastic bag or wrap. If there is printing on the poly wrap, eliminating it and the related handling costs can offset the cost of the heavier cover.

- **Reduce the thickness and weight of packaging.** You'll save shipping costs as McDonald's Corporation did (see Chapter 5).

- **Increase the use of recycled content.** Display the recycled and recyclable symbols on all packaging and papers that qualify. Show the percentage of PCM[27] and PIM[28]. It will inform customers of your environmental message while reminding them to buy recycled and to recycle, too.

- **Influence incoming packaging.** Exert your position with vendors by giving preference to those whose products are in preferred packaging or who switch to reusable containers.

- **Reduce your use of tree-based paper.** Consider buying tree-free paper made of kenaf or other hearty, fast-growing plants.

Reuse

- **Designate central locations for repositories of reusables.** Have employees deposit and retrieve reusable rubber bands, paper clips, binders, file folders, and paper that can be reused for drafts and note paper.

- **Eliminate fax cover sheets.** ReUsables sells re-markable pages in several sizes that can be used over and over again with water-based markers. In fact, if you use small sticky notes on the first page of the document with sender and destination information, you'll save paper at the recipient's fax machine.

- **Reuse packaging.** "Boxes delivered to the mailroom at Kinney Shoe Corporation in Manhattan are reused for shipping, making the purchase of new boxes unnecessary and saving approximately $1,820 a year."[29]

- **Encourage employees to reuse padded envelopes, diskette mailers, and boxes that they receive to send their own outgoing mail and shipments.** So that recipients don't think that you're being cheap, include a note or sticker informing them that your organization minimizes waste and recycles. You'll pass along the idea and reduce your mailing and disposal costs, too.

- **Set up one or two bins for employees to deposit incoming polystyrene peanuts and bubble pack. Reuse these in your own outgoing packages.** You'll reduce waste and costs. Better yet, invest in a large paper shredder for your waste paper and use the shredded paper as padding. That's what businesses used before polystyrene was invented. Both practices will save you the cost of new packaging, avoid disposal costs for incoming materials, and save landfill space.

- **Reuse delivery pallets.** Ask vendors to back-haul their pallets and negotiate that into their contracts. Make the shipping and receiving departments part of the discussions.

As we've seen, more than half of McDonald's suppliers reduced timber demand by 60% and lowered their operating costs by creating a centralized pool of durable reusable wooden pallets. In another case, "IBM was paying $8 a piece for pallet removal, yet when surveyed, all of their vendors said they would be willing to pick the pallets up on the next delivery."[30]

- **Explore the feasibility of using shipping containers made of recycled plastic instead of corrugated.**

- **Support local charities with donations of old furniture and equipment.**

- **Save money by buying used furniture and equipment when appropriate.**

Recycle

- **Plan a facility-wide recycling program.** Start with all types of paper products and corrugated. Contact local recyclers and government agencies for information and recommendations.

- **Work with your recycler or consultant.** Have him/her provide appropriate receptacles throughout the facility for waste paper and other recyclables.

- **Decide whether each employee's work area will have a single can for waste paper, two-cans, or even three receptacles.** The one-can system discourages food at people's desks since all food wastes must be put into a central receptacle; however, it increases the risk of contaminating recyclables when people are too busy, forget, or just don't care. With two cans, employees never have to leave their work areas to properly dispose of anything. Earning the highest reimbursement rates means separating white from mixed paper and may mean three receptacles per employee. Your choice depends on which is best suited to your organization's culture.

- Put larger color-coded bins for recyclables in central locations throughout the facility. Include meeting rooms, the cafeteria, break rooms, at the water cooler, and near lavatories.

- Be sure to have a paper recycling container at every copier, printer, and facsimile machine. Put up signs or posters as reminders.

- Buy Recycled. The U.S. EPA newsletter Re-Usable News reminds us, "If you're not buying goods made from recycled products, you're not recycling."[31] So complete the recycling process by buying recycled goods.

- Target products that are not made of recycled materials, and replace them with recycled equivalents. Pinpoint those with recycled material that can have a higher percentage of post-industrial material. Work with manufacturers and suppliers to achieve higher PCM in existing product lines or replace them with products that already meet this specification.

- Start with disposable papers. These are papers that will be used once, then thrown away or taken out of circulation. Purchase brands with 100% recycled content that includes a large percentage of post-consumer content:
  - Paper towels
  - Toilet tissue
  - Facial tissue
  - Napkins
  - Paper plates and cups

- Make sure that all other paper products have recycled fibers with at least 20% post-consumer content. Focus especially on stationery, envelopes, and copier and printer paper:

- Annual reports
- Business cards
- Carbonless forms
- Cardstock
- Computer greenbar
- Copier
- Corrugated boxes
- Envelopes (even window envelopes)
- Fax paper (even the thermal kind)
- Greeting cards
- Invoices and statements
- Labels
- Legal and steno pads
- Letterhead
- Manuals
- Marketing literature
- Newsletters
- Notepads
- Packaging materials (or shred your own)
- Printing paper
- Promotional materials
- Proxy forms
- Stationery and writing paper

- **For glossy promotional pieces and product brochures, use recycled stock.** It can have the same look and feel as virgin fiber stock. "You might even buy your own waste paper back. HBO saved $40,000 by using its own waste office paper to produce its corporate letterhead."[32]

- **Use only cardboard and packaging made of recycled fibers that are oxygen bleached or unbleached.** Work with vendors to receive their products in only recycled packaging made without chlorine bleaching.

- **Use recycled paper in copiers and printers.** Most of them can handle it with no adjustments. If your equipment can't, leasing office equipment that can do so may be more affordable than buying.

Redesign

- **Look for opportunities to redesign processes so that less waste and pollution are generated and fewer resources are used up.** As stated in Chapter 4, by changing two manufacturing processes, an AT&T plant in Ohio went from being the state's second largest polluter in 1987 to eliminating the offending pollutant altogether with a $210,000 savings.

- **Use environmentally friendly products.** Switch to unbleached products and benign cleaning solutions.

- "Redesign" the office environment as well as manufacturing processes. For example, look closely at reporting procedures. Scrutinize who generates reports, how often, to what detail, and for how many people. Get creative with how things are done now and how they could be done better.

- **Campaign for energy conservation.** Install motion detectors, Energy Star computers, monitors, and office equipment. Post reminders for employees to turn off lights when they're the last ones to leave a room.

Workshop Experience at All Levels

- **Announce the philosophy, purpose, plans, and benefits of the program in writing.** Don't stop there. The written word is too easily filed, recycled, thrown away, or ignored.

- **Conduct all-employee meetings.** Gather in an auditorium or the cafeteria to allow task force and committee leaders to talk to co-workers face-to-face and answer questions.

- **Crusade at department staff meetings.** They're a good place to describe the process and extinguish brush fires from saboteurs who might prefer to see the project go up in smoke.

- **Campaign at manager meetings.** They're an ideal forum to reinforce the company philosophy and mission.

Reinforce the Spoken Word

- Post reminders and progress reports near copiers and printers, on bulletin boards, in cafeterias, and in break rooms. The little bit of paper invested to keep people aware can pay off with big savings. Use electronic mail to update employees on the progress. AT&T Power Systems' monthly meetings have helped maintain momentum.

- Encourage new hires to form good habits from the beginning. Include information in their orientation packets and have Human Resources or their managers discuss the importance of following the program. Naturally, have orientation packets made with post-consumer materials.

Overcome Objections and Obstacles

Not everything will run according to plan. Roadblocks are inevitable, so anticipate them and be prepared to overcome them.

- **Lack of time.** Managers may be concerned that the time employees spend on recycling will mean that they spend less time doing the jobs for which they're paid, especially if employees whine that recycling takes too much time. For this reason, it's imperative to make recycling and waste reduction as easy as possible. Review the section on strategies and remember that separate under-the-desk receptacles for recyclables and trash will avoid the need for people to interrupt their work flow. Centrally located bins at printer stations, copiers, restrooms, and cafeterias will both encourage and remind people to exercise their responsibilities.

- **Lack of motivation.** Human nature guarantees that some people will be indifferent, even hostile toward the changes that must take place. Use creative ideas to spark interest.

  Run a contest to generate suggestions for reducing, reusing, recycling, and redesigning.

  Build incentives for employees who contribute innovative and cost-saving ideas and for those who go above and beyond the call of duty. Get a lot of people involved with a monthly award like the one AT&T Power Systems presents to the department doing the best job.

  Conduct regular events such as Earth Day celebrations. Vendor days will enable employees to learn about recycled products and receive free samples. Tree plantings, beach and community clean-ups, and programs with local schools and neighborhood organizations can be fun while achieving their objectives and spreading the word.

For management, paint a vivid picture of the long-term benefits of taking action and the lasting consequences of not acting. Sometimes negative consequences such as non-compliance penalties, poor employee morale, and bad press are stronger motivators than the positive aspects.

- **Feelings of helplessness.** These may occur at the onset of the program, although they're more likely to happen after the program has been in place for a while. Whenever they occur, several tactics can refuel enthusiasm.

  Set interim goals to measure success along the way. As long as employees see some progress, they'll maintain interest.

  Implement a no-fault approach. Keeping self-esteem high is a better remedy than blaming someone for failures or disappointments.

  Evoke a can-do attitude. Let employees know that they can make a difference, from those at the top who set the direction to those at the bottom who take out the trash.

  Inform employees about what other organizations have done, especially those in the same industry. You'll awaken the competitive instinct and uncover new ideas and methods.

  Publicize company and employee successes, both inside and outside the company. People will get rewards for good deeds that might otherwise go unnoticed.

  Demonstrate concern for people and achievement. Recycling coordinators can work with department managers to spark interest.

- **Lack of compliance.** Roll out a join-the-team campaign. Excite team spirit and pride with posters encouraging compliance from everyone. You might even arrange for a marching band to kick things off.

With those non-conformists who threaten teamwork, instill the sense of adventure and challenge that the world's explorers and pioneers have had. They may eventually become your best leaders for the cause.

- **Lack of personal relevance.** This is probably the single largest obstacle. Many of the world's great salespeople share the premise that while shoppers decide not to buy for logical reasons, they decide to buy for emotional reasons. Sell people on your ideas by helping them understand what's in it for them.

  Influence people with posters and pictures. Depict the consequences of perpetuating wasteful habits and the benefits of changing them will influence a lot of people.

  Develop programs to help employees "own" the project. Give special recognition or monetary rewards to individuals who devise innovative or cost-saving strategies.

  Use some of the proceeds for employee events and programs. Celebrate with free lunches and family picnics.

  Write a regular column in the company newsletter. Recognizing employee and department efforts. Publish featured articles written by employees on relevant issues and recognize them for their contributions.

  Translate problems and solutions into terms that are important to each of the different stakeholders. Use financial and legislative terms for managers and investors; environmental terms for concerned employees and community members. Remember, the key here is to answer the question "what's in it for me?".

- **Lack of momentum.** Enlist, engage, and mobilize everyone. Ask for volunteers who believe in the philosophy and goals of the program. Have each of them contribute to the planning and implementation processes. Set group and individual goals. Establish a way of reporting successes and awarding achievements.

Phase 3: System Implementation and Change—How Can You Ensure You Will Get There?

Some of the strategies may seem redundant. Keep in mind that just as an Olympic athlete must practice relentlessly to win the gold, similar repetition and reinforcement will win a medal for your organization.

Implementing the system and affecting change is the true test of leadership. It will take leaders at all levels of the organization to accomplish the proclaimed goals.

Develop Leadership at All Levels

Start with the executives. Have them kick off the program and proclaim the company's environmental philosophy.

Repeat and renew education campaigns and awards for outstanding efforts.

Build up your organization's image with media coverage, advocacy advertising, and community involvement. Send press releases to local newspapers and community groups. Neighbors and employees will enjoy reading about themselves, and the positive press will help.

Publicize your organization's philosophy. Put the recycled and recyclable logos on everything that qualifies: promotional materials, invoices, and other documents—using paper with at least 20% PCM, naturally. It will be a subliminal reminder to employees while getting the message across to those outside the organization.

Find out which local charities appeal to employees. Consider shelters for the homeless, child development programs, and environmental organizations Recruit workers for a donation committee that decides where some of the profits from recycling will go.

Develop Company Environmental Awareness

Formalize the organization's environmental philosophy. Institute policies, procedures, programs, and structure that are an integral part of the organization. Incorporate them into the employee handbook and new- hire orientation.

Involve groups and individuals. Eliminate sore spots at the department level. Schedule department clean-ups where everyone wears casual clothes to clear out work areas, old files, and storage cabinets together, and to identify resources that can be reused.

Renew people's excitement. Have special events and vendor displays a few times a year: in April for Earth Day; in July to celebrate summer with a park or beach clean-up; in October for Population Week; and in November for Thanksgiving. December is a good time to donate aluminum can reimbursements to charity or run a drive for reusable items like clothing.

Inform employees of other conservation efforts outside of work. Empower them to implement conservation on their own.

Budget money to move efforts forward. After the plan is running smoothly, you'll still need resources to keep it going. This may warrant a committee, on-going education, or simply periodic audits and assessments.

Phase 4: Evaluation, Adjustment, and Revitalization— Where Do We Go from Here?

Once the strategies are in place, you'll need to monitor, modify, and fine-tune them. Do this at least once every quarter.

Evaluate Objectives and Achievements

Conduct accurate measurements. Update your Waste Reduction Goal Worksheet regularly. Assess the progress, setbacks, and priorities. Using your Monthly Cost Savings Worksheet and Waste Reduction Goal Worksheet, base the measurements on performance results versus promises.

Be pragmatic and quantify the results. To what degree have quantifiable goals been achieved? Have costs been reduced? By how much? What new revenues have been generated?

Identify cultural results. What new habits and philosophies have been adopted?

Designate and evaluate areas for improvement. These may include adding or upgrading related equipment, training staff on electronic mail, reducing the use of disposable items, or maybe just renewing your "internal marketing" efforts.

Report

Publish quarterly reports to "stakeholders." Target employees, the finance department, all levels of management, and community members. The reports will hold the task force, committee members, departments, and individuals accountable for achieving results.

Adjust

Acknowledge achievements and shortfalls. Then revise weak areas. Look to vendors and outside consultants for assistance if necessary.

Obtain commitment from management to continue the program as is or to modify it for greater success. Discuss performance objectives for managers, employee bonuses, additional profit-sharing contributions, and shareholder dividends.

Set or modify company-wide waste minimization standards. Raise goals for the recycling/recovery rate, reducing consumption and costs, informing stakeholders, and engaging employees.

Revitalize

Sustain momentum with employees. If periodic department and all-employee meetings, employee activities, awards, and charitable contributions have been working, continue them. If not, explore alternatives. Poll employees for their preferences.

Sponsor free community seminars on environmental issues. This is an opportunity to get positive press and goodwill while influencing others.

Publish and distribute free booklets on conservation and environmental protection programs that employees can implement at home. One of the driving forces for many environmental actions is to provide our children and grandchildren with a wholesome, healthy world for them to grow. Bringing good habits home means passing them along to the next generation and ensuring real long term results.

Extend the program to other areas of the company. Implement the program in satellite offices around the country. Determining milestones may take longer because of the time involved in researching the local markets–vendors, rates, regulations, etc. Implementing the program may be the easier part of the project because satellite facilities are usually smaller and involve fewer people. If they're very small, you may decide to involve neighboring businesses or other tenants in the building, creating the "critical mass" necessary to make the program worthwhile. Refer to the Atlanta hub example in Chapter 4 for ideas.

Expand efforts to worldwide operations. Remember that resources, traditions, and customs will vary from country to country, so be prepared for a longer ramp-up period. Work with managers and employees to provide the tools, systems, and initiative to achieve success.

Review the Process

The Council on the Environment of New York City Waste Prevention & Recycling Service encapsulates many of the most effective activities, shown in Figure 10-7:

- Create an Environmental Focus Group with a waste-reduction agenda.
- Take a look at one day's waste to get an accurate picture of what is being discarded.
- Speak with your carter to learn how much waste is picked up and what this is costing you.
- Speak with departments—Facilities, Purchasing, Cafeteria, Art Departments, Marketing, Reproduction, Printing, etc.— about the procedures within each department that produce waste; discuss alternatives.
- Make a cost analysis of the benefits of a switch to waste-saving practices, such as durable dishware in a cafeteria, double-sided copying, or donating discards.
- Disseminate the results, asking for management support and employee cooperation.
- Make waste reduction part of your organization's policy statement. Ask for floor and/or department monitors.
- Communicate company policy to all vendors and ask their cooperation by reducing packaging, instituting reusable systems when possible, and back-hauling all pallets.
- Meet with monitors regularly to measure the results of the initiatives implemented and the savings realized; publicize the results.

Figure 10-7. Eight steps guaranteed to reduce waste.[33]

ENDNOTES

1. Robert F. Allen and Charlotte Kraft, *Organizational Unconscious: How to Create the Corporate Culture You Want and Need* (Englewood Cliffs, NJ: Prentice-Hall, 1982).

2. James A. F. Stoner and R. Edward Freeman, *Management*, 4th ed. (Englewood Cliffs, NJ: Prentice Hall, 1989), p. xxii.

3. Harold Koontz, Cyril O'Donnell, and Heinz Weihrich, *Management: A Book of Readings*, 5th ed. (New York, NY: McGraw Hill, 1980), pp.vii–x.

4. Varies depending on the compaction ratio.

5. Implementation plans and worksheets modeled after those in *Successful Business Recycling* (Basking Ridge, NJ: AT&T) and *Business Recycling Manual* (New York, NY: Inform, Inc. and Recourse Systems, Inc., 1991).

6. Include trucking fees, dumping or tipping fees, and taxes.

7. Based on compaction ratio.

8. From Figure 10-3, Waste Audit Worksheet.

9. Ibid.

10. The sudden jump in prices between 1993 and 1995 is due to several factors, including an improved economy, higher demand for paper goods, and newer mill technology. Even as this book is going to press, rates are dropping dramatically.

11. Paper stock prices published in various issues of *Fibre Market News*, Cleveland, OH, as quoted in an interview with Dan Sandoval, Editor, March 27, 1995. Prices are not exact, but rather serve as a guideline of rates for different grades in various markets.

12. From Figure 10-2, Disposal Cost Worksheet.

13. *Waste Reduction: A Guide for New York Businesses* (New York, NY: Council on the Environment of New York City Waste Prevention & Recycling Service), p. 5.

14. Ibid.

15. Ibid.

16. Ibid.

17. *Fifty Ways Your McDonald's Can Help the Environment* (Oak Brook, IL: McDonald's Corporation Environmental Affairs, 1992), p. 9.

18. *Waste Reduction: A Guide for New York Businesses*, p. 4.

19. Ibid.

20. Ibid.

21. Ibid., pp. 4-5.

22. Ibid., p. 5.

23. Ibid.

24. Ibid.

25. Ibid.

26. *Waste Reduction: A Guide for New York Businesses*, p. 4.

27. Post-consumer material.

28. Post-industrial (pre-consumer) material.

29. *Waste Reduction: A Guide for New York Businesses*, p. 4.

30. Ibid.

31. *The Warmer Bulletin, Number 43*, (Tonbridge, Kent, UK: The World Resource Foundation, November 1994), p. 8.

32. *Waste Reduction: A Guide for New York Businesses*, p. 5.

33. Ibid., p. 7.

PART IV:

BEYOND THE PAPER MESS

CHAPTER 11

WHERE ELSE CAN WE GO FROM HERE?

Reducing consumption of virgin pulp and diverting paper products from the waste stream may be the two best ways to diminish the waste crisis. But they're only the beginning in the long fight to save the environment. There are many other things we can do.

"Job displacement which results from recycling, weighted against employment and efficiency gains in environmental protection, suggest that we are increasingly better off as we expand the re-use and recycling of 'waste' materials."[1]

Re-evaluate Office Supplies and Practices

Recycle your used toner cartridges. Buy recycled cartridges, too.

Increase the use of reusable office supplies. Get employees to use mechanical pencils. Provide them with refills and extra erasers.

Change your correction fluid. Switch to low-toxic correction fluid (Gillette reformulated its Liquid Paper[2]) or correction tape.

Reduce Packaging Waste

Buy in bulk. If you have the storage room, reduce the amount of packaging waste you generate by buying in larger quantities. You'll buy less often, which reduces purchasing, personnel and processing costs, and lets you take advantage of bulk discounts.

Buy products with less packaging. Most packaging is designed to withstand far more pressure and abuse than it will normally undergo. Suppliers will respond to your requests because it will save them money in lower raw materials and transportation costs.

Reduce Organic Waste

Practice composting. This is an area where some McDonald's stores have already achieved admirable results. They've shown that food establishments can work with local farmers or wholesale nurseries to make the most out of food and other organic wastes.

Reduce Toxins

"Inquire about what chemicals are in a product before purchasing, and refuse furnishings that can be toxic."[3]

Boycott fiberboard. Although cheap, fiberboard is made with glues containing formaldehyde. "Found in many office products including particle-board, fiberboard, and plywood furniture and panelling, carpeting, glues, and upholstery and drapery fabrics, this chemical gives off a gas associated with adverse respiratory effects and is also a potential carcinogen."[4]

Instead, choose a high-density plywood with a very low percentage of glue. For internal construction and furniture, use panels made of 100% recycled paper, such as products from Gridcore Systems International. The panels are made with non-toxic glue, and there are no harmful emissions in the manufacturing process.

Re-evaluate your inks. Switch from chemical-based inks and pigments to soy or other vegetable-based products. They are far less toxic to manufacture and to recycle. Standardize on environmentally friendly dyes and art supplies, too.

"Every ink color has toxic substances in it but some contain higher amounts of toxicity than others....Avoid the most toxic ink colors...by working with your printer in choosing barium-free PMS numbers or asking your printer to reformulate the ink to take out the barium." Avoid "copper metallics; most other metallics...are made from aluminum flakes and pigment and are not a problem."[5]

Phase out glossy paper. "Consider alternatives to the glossy cadmium-laden color stock often used for promotional (materials).... If it gets incinerated, those chemicals (used to manufacture it) go into the air."[6] Besides, not all recycling programs can handle glossy paper.

Reduce Vehicle Emissions

Share the ride. Our thirst for energy and the fumes our fossil fuel vehicles emit are among the biggest threats to the planet. Help cut down on harmful emissions by arranging or increasing car and van pooling and public transportation. Car and van pooling will cut smog while saving employees money, building friendships, and making the commute more enjoyable.

Reduce Energy Consumption

Turn off lights. Install motion detectors in offices, conference rooms, and other common areas. If that's not feasible, put stickers on the light switchplates reminding the last one out to turn off the lights.

Evaluate air conditioning systems and lighting fixtures. Run a break-even analysis to see how soon switching to electronic ballasts for fluorescent lighting will yield a return on investment.

Choose Energy Star certified equipment for new purchases. They will power down when not in use. (Check the factory settings to make sure that power-down mode is on.)

Work with suppliers. Contact your gas and electricity providers for a complete energy analysis and recommendations.

Reduce Water Consumption

Don't be a drip. Turn off the sprinklers in cool or rainy weather. Install low-flow toilets. Repair all leaking plumbing, and don't leave the water running.

Use Reusables

Encourage the use of reusable mugs. Issue reusable mugs to new employees. Post reminders near beverage dispensers and give a price break to those shun throw-away cups.

Standardize on reusable shipping containers and pallets. Look for other areas where reusables can replace disposables.

Donate Reusables

Donate durable goods including office furniture and equipment. They "make up 11% of our waste stream. Though much of this material can still be used, ... repaired, or (salvaged for) valuable scrap materials, only a small percentage of it ever finds its way to a second owner."[7] Donations to groups that can use these items are often tax-deductible and many groups will pick up the materials. If your organization exceeds the maximum amount allowed for tax deductions, it will still reduce disposal costs and keep these bulky items out of the waste stream.

Donate magnetic media. "EcoMedia is a national recycling organization. We provide services and support to people with severe disabilities through the recycling of magnetic media....Our primary work is reprocessing and recycling or dismantling for parts and material content." The organization accepts new, used, or rejected videotapes, audiotapes, mastering tapes, floppy disks–magnetic media of all kinds. Those that are in good condition are donated to nonprofit organizations. Unusable plastics and magnetic material are recycled. Confidentiality and degaussing are an integral part of the process. Donations are tax deductible, too.

Recycle Plastics

Plastics recycling has come a long way in the last ten years. Recycled plastics can come back in many different forms, as shown in Table 11-1. In addition, old milk containers and plastic bags can be made into plastic lumber. Clothing can also be made from recycled plastic.

Table 11-1.   Forms of Some Recycled Plastics[8]

Type of Plastic	Recycled Into...
PET	Bakery and deli trays, carpets, fiberfill, geotextiles, and soft drink bottles
HDPE	Agricultural pipe, bags, bottles for laundry products, recycling bins, soft drink bottle base cups, and motor oil bottles
Vinyl	Fencing, non-food bottles, and pipe
LDPE	Bags and film (wrap)
PP	Auto parts, carpets, geotextiles, and industrial fibers
PS	Cafeteria trays, insulation board, office accessories, toys and accessories, video cassettes and cases

Recycle Aluminum

Using reclaimed aluminum saves about 95% of the energy needed to produce virgin aluminum from bauxite, which is very harsh on the environment. Recycling one aluminum can saves enough energy to run a television for about three hours.[9]

Aluminum can recycling in the United States is estimated at 33%. This is almost double what it was in the mid-1970s, and higher than Britain's 1993 aluminum can recycling rate of 21%.

One study done several years ago revealed that approximately 30,000 jobs in the United States resulted exclusively from aluminum recycling. That number increased to approximately 3 million people nationwide involved in the collection of aluminum scrap and beverage cans.

Recycle Polystyrene

At one time recycling polystyrene was neither practical nor cost effective. It was particularly problematic with food and beverage containers. That is no longer true. Polystyrene is recycled into insulation, packing material, and many other non-food-related goods.

Redesign

The importance of cradle-to-grave product planning cannot be emphasized enough. It is the best way companies can reduce toxins, consume less energy, and generate less waste.

Germany is perhaps at the forefront of driving this concept forward with legislation and innovation. Yet, even Germany has a long way to go to institute standards and methodologies that make it both easy and cost effective for consumers and manufacturers.

In 1995, Siemens Nixdorf in Germany was

> awarded the first Blue Angel ecolabel for a computer. The award reflects the fact that the computer was specifically designed to be reused or recycled....Matsushita has developed a television which can be taken apart by the removal of four screws. Hitachi has developed a washing machine with a tank made of stainless steel, because it will be more recyclable than plastic. Mistubishi has reduced the number of different models on offer, and minimized the number of different components.[10]

The United States can learn a lot from other nations about the recovery and recycling of electronic equipment. Fortunately, several electronics recycling companies are already in operation in the United States. When possible, they refurbish and resell fully functional equipment. If resale is not feasible, they redeploy the components that are not obsolete and still usable, such as hard drives, memory modules, and add-in boards. They sell off the remains to salvage and recycling firms. Scrap metal recyclers melt down the metal chassis and the precious metals on many of the connectors and components. Some plastics can be recycled as well.

As with everything else, the more research and competition increases, the more affordable solutions will become available. Ultimately it's up to us, the consumers, to create the demand for the best alternatives.

ENDNOTES

1. "Winners and Losers: Employment Impact of Recycling," *Biocycle: Journal of Composting and Recycling* (Emmaus, PA: J. G. Press, Inc., March 1988), p. 46.

2. Bette K. Fishbein and Caroline Gelb, *Making Less Garbage: A Planning Guide for Communities* (New York, NY: INFORM, Inc., 1992), pp. 144–145.

3. *Waste Reduction: A Guide for New York Businesses* (New York, NY: Council on the Environment of New York City Waste Prevention & Recycling Service), p. 10.

4. Ibid.

5. Ibid.

6. Ibid., pp. 10–11.

7. Ibid., p. 6.

8. *Plastics in Perspective: Answers to Your Questions About Plastics in the Environment,* (Washington, DC: American Plastics Council, 1993), p. 12.

9. *Reynolds Makes and Takes the Wrap,* (Richmond, VA: Reynolds Metals Company, Consumer Products Division), p. 4, and *Wrap with Renolds* (Richmond, VA: Reynolds Metals Company, Consumer Products Division), p. 2.

10. *Warmer Bulletin, Number 44* (Tonbridge, Kent, UK: The World Resource Foundation, February 1995), p. 13.

CHAPTER 12

NEW BEGINNINGS

While paper and board recycling is making good progress, it does face some problems. The difference between under-supply and over-supply of recyclable material can be very narrow. Thus a decline in demand for new paper and board—the inevitable result of lower economic activity—can mean frustration for collectors who will include charities, local community groups and schools....It is clear that paper recy-cling can only be made to work if collection and consumption can be coordinated.[1]

More mills than ever are on-line with the latest technology. Companies and individuals are very much aware of the need to recycle, buy recycled, and reduce consumption in the first place. As we address the challenge of re-engineering how the industrial world does things, we must also turn our attention to the developing nations. How do we keep them from making the same mistakes without keeping them from having the standard of living that the industrial nations have come to enjoy?

We can stop exporting dangerous substances that are illegal in many industrialized nations but are still allowed in other countries. We can help them incorporate cradle-to-grave product planning. We can provide them with newer technologies that may be a little more costly in the beginning but that have at least as good a return with less environmental impact in the long run.

In areas with few or no telephone services, perhaps wireless communication is a better solution than using cables and fiber. Maybe cogeneration and solar panels on buildings can abate the need for coal burning or nuclear power plants.

When U.S. companies go into or expand in foreign markets, they must take responsibility for what they bring. Soft drinks manufacturers can introduce and be responsible for aluminum can and plastic bottle recycling. Computer and office machine manufacturers can choose to sell only newer technology that is energy efficient and easily recycled at the end of its life. When our own energy guzzling office machines are obsolete, we can take them out of service altogether and lessen the strain on energy sources.

As new generations grow up with technology, we must instill in them the values and methodologies that will allow them to pass on a sustainable environment to their children.

We've come far in mending our wasteful ways. We have far to go, and we can all get there by working together to find viable, effective solutions.

ENDNOTES

1. *The Warmer Bulletin, Number 43, Information Sheet: Paper Making & Recycling* (Tonbridge, Kent, UK: The World Resource Foundation, November 1994), p. 4.

APPENDIX A

RECYCLED PAPER MANUFACTURERS

Appleton Papers
Appleton, WI
(800) 558-8390

Arjo Wiggins USA
Greenwich, CT
(203) 622-4503

Badger Paper Mills
Pesthigo, WI
(715) 582-4551

Beckett Paper Co.
Hamilton, OH
(800) 543-1188

Boise Cascade
Portland, OR
(503) 790-9419

Byron Weston Co.
Dalton, MA
(413) 684-1234

Champion Int'l Corp.
Stamford, CT
(203) 358-7000

Consolidated Papers, Inc.
Chicago, IL
(312) 781-0200

CPM, Inc.
Claremont, NH
(603) 542-2592

Cross Pointe Paper Corp.
St. Paul, MN
(612) 644-3644

Custom Papers Group
Richmond, VA
(804) 649-4463

Daiei Papers
Elmwood Park, NJ
(201) 794-3711

Domtar Fine Papers
Montreal, QUE
(514) 848-5649

Eastern Fine Paper, Inc.
Brewer, ME
(207) 989-7070

E.B. Eddy Forest Products
Ottawa, ONT
(800) 267-9971

Ecusta
Pisgah Forest, NC
(704) 877-2211

Esleeck Manufacturing Co.
Turners Falls, MA
(413) 863-4326

Finch, Pruyn & Co.
Glenn Falls, NY
(518) 793-2541

Fletcher Paper Co.
Alpena, MI
(800) 634-3158

Fox River Paper Co.
Appleton, WI
(800) 558-8327

Fraser Paper Ltd.
Stamford, CT
(203) 359-2544

French Paper Co.
Niles, MI
(616) 683-1100

George Whiting Paper
Menasha, WI
(414) 722-3351

Georgia Pacific
Atlanta, GA
(404) 220-8365

Gilbert Paper
Menasha, WI
(414) 722-7721

Hammermill Papers
Memphis, TN
(800) 633-6369

Howard Paper
Appleton, WI
(414) 733-7341

International Paper
New York, NY
(800) 541-4141

Island Paper Mills
New Westminster, BC
(604) 527-2618

James River
Communcation Papers
Oakland, CA
(800) 932-4888

Lyons Falls Pulp & Paper
Crystal Lake, IL
(815) 455-0981

MACtac
Stow, OH
(216) 688-1111

Mead Paper
Chillicothe, OH
(614) 772-3111

Mohawk Paper Mills, Inc.
Cohoes, NY
(518) 237-1740

Monadnock Paper Mills
Bennington, NH
(603) 588-3311

Neenah Paper
Roswell, GA
(800) 241-3405

Niagara of Wisconsin
Paper Corp.
Niagara, WI
(800) 826-0431

P.H. Glatfelter Co.
Spring Grove, PA
(717) 225-4711

Potlatch Corp.
Cloquet, MN
(218) 879-2300

Premium Papers Marketing
Merrick, NY
(516) 378-8832

Provincial Paper
Fairfield, OH
(513) 858-3649

Repap Sales Corp.
Stamford, CT
(203) 353-3333

Riverside Paper Co.
Appleton, WI
(414) 749-2200

Rolland, Inc.
Montreal, QUE
(514) 569-3909

S.D. Warren
Boston, MA
(617) 423-7300

Scheufelen N. America
Roslyn Heights, NY
(516) 621-2470

Seaman Paper Co.
Otter River, MA
(508) 939-5356

Simpson Paper Co.
Seattle, WA
(206) 224-5843

Specialty Paperboard,Inc.
Brattleboro, VT
(802) 257-0365

Springhill Paper
Memphis, TN
(800) 541-4141

Stora Papyrus
Newton Falls, NY
(800) 448-8900

Strathmore Paper Co.
Westfield, MA
(800) 628-8816

Union Camp Corp.
Franklin, VA
(804) 569-4321

Unisource
Wayne, PA
(215) 296-4470

Wausau Paper Mills
Brokaw, WI
(800) 950-9767

Westvaco Corp.
New York, NY
(212) 688-5000

Weyerhaeuser Paper
Wayne, PA
(610) 251-9220

Zanders, USA
Wayne, NJ
(201) 305-1990

APPENDIX B

The information in this section is a partial listing of the Directory of U.S. Paper and Paperboard Mills Consuming Recovered Paper as published in the October 1994 edition of PaperMatcher. It provides a state-by-state listing of pulp, paper, and paperboard mills that utilize recovered paper as a raw material and the types of recovered paper handled by these mills. These mills may work on potential markets for recovered paper directly with generators or through local recovered paper dealers or brokers.

For a copy of PaperMatcher with a complete listing of names, addresses, and telephone numbers, please contact the American Forest & Paper Association, Inc., 1111 19th Street, Suite 800, Washington, DC, 20036, (800) 878-8878.

The paper groupsing utilized by these mills areare defined as follows:

C   Corrugated: old containers, Kraft paper and bags, carrier stock, clippings.

MP   Mixed Paper: including magazines, catalogs, telephone directories, tissue paper, recycled boxboard cuttings.

PS   Pulp Substitutes: Print-free grades of bleached chemical office and computer papers.

HG   High-Grade De-inking: de-inked printed grades of bleached chemical office and computer papers.

N   Newspapers: including old newspapers, special, oversize, and white blank newsprint, groundwood.

State	City	Company Name	Type of Recovered Paper
AL	Albertville	Keyes Fibre Co.	PS
	Anniston	National Gypsum Co.	MP, N, C
	Clairborne	Alabama Newsprint	MP, N
	Coosa Pines	Kimberly-Clark Corp.	MP, N
	Courtland	Champion International Corp.	PS
	Mahrt	Mead Coated Board, Inc.	C
	Mobile	Armstrong World Industries	MP, N
		GAF Corp.	MP, C
		Newark Boxboard Co.	MP, N, C, PS
		Scott Paper Co.	PS
	Pennington	James River Corp.	PS
	Pine Hill	MacMillan Bloedel, Inc.	C
	Prattville	Union Camp Corp.	MP. C
	Selma	International Paper Co. (Hammermill)	MP, PS
	Stevenson	Mead Corp.	MP, C
AZ	Flagstaff	Orchids Paper Products Co.	HG
	Snowflake	Stone Container Corp.	MP, N, C, PS
AR	Ashdown	Georgia Pacific Corp.	PS
	Camden	Celotex Corp.	N
		International Paper Co.	N
	Crossett	Georgia Pacific Corp.	PS
	Morrilton	Green Bay Packaging, Inc.	N, C, PS, HG
	Pine Bluff	Gaylord Container Corp.	N, C
CA	Anderson	Simpson Paper	PS
	Antioch	Gaylord Container Corp.	C
	City of Industry	Packaging Company of California	N
		Sonoco Products Co.	MP, N, C
	Fontana	Fontana Paper Mills, Inc.	MP, N, C
	Fullerton	Kimberly-Clark Corp.	PS
	Hollister	Leatherback Industries, Inc.	MP, C
	La Verne	Paper-Pak Products, Inc.	PS< HG
	Los Angeles	Los Angeles Paper Box & Board Mills	N, C, PS
		Newark Pacific Paperboard Corp.	MP, N, C, PS
	Newark	Inland Container Corp.	C
	Ontario	Inland Container Corp.	C
	Oxnard	Procter & Gamble Paper Products	PS
	Pomona	Smurfit Newsprint	N
		Sierra Tissue, Inc.	MP, N
		Simpson Paper Co.	PS, HG
	Port Hueneme	Willamette Industries, Inc.	C
	Red Bluff	Packaging Company of America	MP, N
	Ripon	Simpson Paper Co.	PS
	Sacramento	Keyes Fibre Co.	N, PS

State	City	Company Name	Type of Recovered Paper
	San Leandro	Domtar Gypsum	MP, N, C, PS
	Santa Clara	California Paperboard Corp.	MP, N, C, PS
		Jefferson Smurfit Corp.	MP, N, C, PS
	Santa Fe Springs	Specialty Paper Mills, Inc.	MP, N, C, PS
	South Gate	USG Industries Co.	MP, N, C
		Lunday-Thagard Co.	MP, N, C
	Stockton	Newark Boxboard	MP, N, C, PS
	Vernon	Domtar Gypsum	MP, N, C, PS
		Jefferson Smurfit Corp.	MP, N, C, PS, HG
		Pabco Paper Products, Inc.	MP, N, C
CO	Denver	Republic Gypsum	MP, N, C, HG
CT	Manchester	Lydall, Inc.	MP, N, C
	Montville	Rand-Whitney Paperboard Corp.	MP, N, C, PS, HG
	New Haven	Simkins Industries	N, C, PS
	New Milford	Kimberly-Clark Corp.	PS
	Uncasville	Stone Container Corp.	C
	Versailles	Federal Paperboard Co., Inc.	MP, N, C, PS, HG
DE	Newark	James River Corp.	PS
FL	Fernandina Beach	Jefferson Smurfit Corp.	C
	Hialeah	Atlas Paper Mills	PS, HG
	Jacksonville	Jefferson Smurfit Corp.	C
		Seminole Kraft	N, C
		USG Corp.	MP, N, C, PS
	Palatka	Georgia Pacific Corp.	PS
	Panama City	Stone Container Corp.	C
	Pensacola	Champion International Corp.	PS
	Port St. Joe	Saint Joe Paper	C
GA	Albany	Procter & Gamble Paper Products	PS
	Atlanta	Sonoco Products Co.	MP, N, C
	Augusta	Augusta Newsprint Co.	HG
		Federal Paper Board Co., Inc.	HG
		Deerfield Specialty Papers	HG
	Austell	Austell Boxboard Corp.	MP, N, C
		Caraustar Industries, Inc.	MP, N, C, PS
	Cedar Springs	Georgia Pacific Corp.	MP, C
	Cedartown	Jefferson Smurfit Corp.	MP, C
	Conyers	Pratt Industries	MP, N, C
	Dublin	Southeast Paper Manufacturing Co.	N
	Macon	Armstrong World Industries	MP, N
		Riverwood International Corp.	MP, C
		Packaging Corp. of America	N, PS
	Pt. Wentworth	Stone Container Corp.	N, C
	Riceboro	Interstate Container Corp.	C
	Rincon	Fort Howard Paper Co.	MP, N, C, PS, HG

State	City	Company Name	Type of Recovered Paper
	Savannah	Union Camp Corp.	MP, N, C
	St. Mary's	Gilman Paper Co.	PS
ID	Lewiston	Potlatch Corp.	PS
IL	Alsip	FSC Paper	N, HG
	Alton	Jefferson Smurfit Corp.	MP, C
	Aurora	The Davey Co.	MP, PS
	Chicago	Chicago Paperboard Corp.	MP, N, C
	Joliet	Ivex Corp.	MP, N, C, PS
		Manville Roofing Div.	N
	Pekin	Quaker Oats Co.	N, C, HG
	Peoria	Ivex Corp.	N, C, PS
	Quincy	Celotex Corp.	N, C, HG
	Rockton	Sonoco Products Co.	MP, N, C
	Taylorville	Georgia Pacific Corp.	PS, HG
IN	Brownstown	Kieffer Paper Mills, Inc.	MP, C, PS
	Carthage	Jefferson Smurfit Corp.	C
	Eaton	Rock-Tenn Co.	MP, N, C
	Gary	Georgia Pacific Corp.	MP, N, C
	Griffith	Packaging Corp. of America	N
	Hammond	Keyes Fiber Co.	N,C
	Hartford City	Visy Recycling, Inc.	MP, N, C
	Indianapolis	Simkins Industries, Inc.	N, C, PS
	Lafayette	Jefferson Smurfit Corp.	MP, N, C
	Newport	Temple-Inland, Inc.	C
	Terre Haute	The Weston Paper & Manufacturing Co.	C
	Wabash	Jefferson Smurfit Corp.	MP, N, C, PS
IA	Fort Madison	Consolidated Packaging Corp.	N, C
	Tama	Packaging Corp. of America	MP, N, C, PS, HG
KS	Hutchinson	Republic Paperboard Co.	MP, N, C, PS, HG
	Phillipsburg	Tamko Asphalt Products of Kansas	MP, C
KY	Hawesville	Willamette Industries, Inc.	C
	Maysville	Temple-Inland, Inc.	C
LA	Bogalusa	Gaylord Container Corp.	C
	Campti	Willamette Industries, Inc.	C
	Deridder	Boise Cascade Corp.	C
	Hodge	Stone Container Corp.	N, C, PS
	Lockport	Nicolaus Paper	PS
	Marrero	Celotex Corp.	N
	Port Hudson	Georgia Pacific Corp.	PS
	Shreveport	G. S. Roofing Products Co., Inc.	MP, C
	St. Francisville	James River Corp.	MP, HG
	West Monroe	Riverwood International Corp.	MP, C
ME	Augusta	Statler Tissue Co.	C, PS, HG
	Brewer	Eastern Fine Paper, Inc.	PS

State	City	Company Name	Type of Recovered Paper
	Bucksport	Champion International Corp.	PS, HG
	East Millinocket	Bowater, Inc.	MP, N
	Gardiner	Yorktowne Paper Mills of Maine	MP, N, C, PS
	Jay	Otis Specialty Papers, Inc.	PS
	Lincoln	Lincoln Pulp & Paper, Inc.	PS
	Lisbon Falls	Wood Fiber Industries	N
	Madawaska	Fraser Paper Ltd.	N
	Old Town	James River Corp.	PS
	Skowhegan	Scott Paper Co.	PS
	Waterville	Keyes Fiber Co.	N, PS
	Westbrook	Molded Fiber	N, C
		S. D. Warren Co.	MP, PS
	Winslow	Scott Paper Co.	PS, HG
	Woodland	Georgia Pacific Corp.	PS, HG
MD	Baltimore	Chesapeake Corp.	MP, N, PS
	Ilchester	Simkins Industries	MP, N, C, PS
	Williamsport	Maryland Paper	MP, N, C
MA	Adams	James River Corp.	PS
	Baldwinville	American Tissue Mills of MA, Inc.	MP, N, C, PS, HG
	Boston	Perkit Folding Box Corp.	MP, N, C, PS
		Patriot Paper Corp.	PS, HG
	Dalton	Crane & Co.	PS
	E. Pepperel	Merrimac Paper Co., Inc.	MP
	Erving	Erving Paper Mills	PS, HG
	Fitchburg	James River Corp.	MP
	Haverhill	Haverhill Paperboard Corp.	MP, N, C, PS
	Holyoke	Sonoco Products Co.	MP, N, C, PS
	Housatonic	Rising Paper Co.	PS
	Lawrence	Newark Atlantic Paperboard Corp.	MP, N, C, PS
		Merrimac Paper Co., Inc.	MP, N
	Lee	Kimberly-Clark Corp.	PS
	Natick	Newark Boxboard	MP, N, C
	Otter River	Seaman Paper Co. of MA, Inc.	PS
	Russell	Westfield River Paper Co.	PS
	Turners Falls	Esleek Manufacturing Co., Inc.	PS
	Westfield	International Paper Co.	PS
	W. Springfield	Decorative Specialties International, Inc.	N, PS
MI	Alpena	Fletcher Paper Co.	PS
	Battle Creek	Michigan Paperboard Co.	MP, N, C, PS, HG
		Waldorf Corp.	MP, N, C, PS
	Constantine	Simplex Products	MP, N, C, PS
	Detroit	Ivex Corp.	PS
	Escanaba	Mead Corp.	PS, HG
	Filer City	Packaging Corp. of America	MP. C

State	City	Company Name	Type of Recovered Paper
	Kalamazoo	Georgia Pacific Corp.	MP, N, PS, HG
		James River Corp.	MP, N, C, PS
	L'Anse	Celotex Corp.	MP, N, C, PS
	Manistique	Manistique Papers	MP, N
	Menominee	Menominee Paper Co., Inc.	C, PS
	Monroe	Monroe Paper Co.	MP, N, C
		Jefferson Smurfit Corp.	MP, N, C, PS,HG
	Muskegon	Scott Paper Co.	PS
	Niles	French Paper Co.	HG
		Simplicity Pattern Co., Inc.	MP, PS
	Ontonagon	Stone Container Corp.	C
	Otsego	Menasha Corp.	C
		Rock-Tenn Co.	MP, N, C
	Palmyra	Big M Paperboard, Inc.	MP, N, C, PS
	Parchment	James River Corp.	C
	Plainwell	Simpson Paper Co.	PS
	Port Huron	Eb Eddy Paper, Inc.	PS
		James River Corp.	PS
	Rochester	James River Corp.	C
	Rockford	Midwest Folding Carton	N, C, PS, HG
	Vicksburg	Simpson Paper Co.	PS
	Watervliet	Fletcher Paper Co.	PS, HG
	West Pigeon	White Pigeon Paper Co.	MP, N, C, PS
	Ypsilanti	James River Corp.	MP
MN	Brainerd	Potlatch	PS
	Cloquet	Potlatch	PS
	Duluth	Pentair, Inc.	PS
	Grand Rapids	Blandin Paper Co.	PS, HG
	International Falls	Boise Cascade Corp.	PS
	Little Falls	Hennepin Paper Co.	MP, PS
	Sartell	Champion International Corp.	MP
	Shakopee	Certain-Teed Corp.	MP, C
	St. Paul	Waldorf Corp. (Boxboard)	MP, N, C, PS
		Waldorf Corp. (Corrugated)	MP, C
MS	Columbus	Weyerheauser	MP, HG
	Meridian	Celotex Corp.	MP, N, C, PS
		Atlas Roofing Corp.	MP, N, C
	Monticello	Georgia Pacific Corp.	C
	Natchez	Manville Corp.	N
	Pickens	Burrows Paper Corp.	PS
	Vicksburg	International Paper Co.	C
MO	Boonville	Huebert Fibreboard, Inc.	MP, N
	Joplin	Tamko Asphalt Productss, Inc.	MP, N, C
	N. Kansas City	USG Corp.	MP, N, C

State	City	Company Name	Type of Recovered Paper
MT	Missoula	Stone Container Corp.	MP, C
NH	Berlin	James River Corp.	MP
	Claremont	APC Corp.	C, PS, HG
	Groveton	Groveton Paperboard	C
		James River Corp.	PS
	Hinsdale	Ashuelot Paper Co.	PS
		GE Robertson & Co.	MP, N
		Paper Service Mills	PS
	Penacook	Penacook Fibre	N, C
	Tilton	Quin-T Corp.	N
	W. Hopkinton	Papertech Corp.	MP, N, C, PS
NJ	Camden	Camden Paperboard Corp.	MP, N, C
	Clark	USG Corp.	MP, N, C, PS
	Clifton	Recycled Paperboard, Inc.	MP, N, C
	Delair	Georgia Pacific Corp.	N, C, PS
	Elmwood Park	Marcal Paper Mills, Inc.	PS, HG
	Garfield	Garden State Paper Co., Inc.	N
	Garwood	Mafcote Industries	MP, N, C
	Hughesville	James River Corp.	PS
	Jersey City	Davey Co.	MP, PS
	Milford	James River Corp.	PS
	Newark	Newark Boxboard	MP, N, C
	Paterson	Paper Board Specialties, Inc.	MP, N, C, PS
	Ridgefield Park	Simkins Industries	N, C, PS
	Warren Glen	Custom Papers Group, Inc.	PS, HG
	West Trenton	Homasote Co.	N
NM	Albuquerque	Leatherback Industries	MP, C
	Prewitt	McKinley Paper	C
NY	Amsterdam	Sonoco Products Co.	MP, N,
	Beaver Falls	Specialty Paperboard, Inc.	MP, N, C, PS
		Specialty Paperboard, Inc. (Lewis)	MP, N, PS
	Brownville	Brownville Specialty Paper Products, Inc.	MP, N, C, PS
	Carthage	Climax Manufacturing Co.	MP, N, C, PS, HG
		James River Corp.	MP, N, C, PS
	Castleton-Hudson	Fort Orange Paper Co.	MP, N, C, HG
	Chatham	Yorktowne Paper Mills, Inc.	MP, N, C, PS, HG
	Cohoes	Mowhawk Paper Mills, Inc.	MP, PS, HG
	Corinth	International Paper Co.	PS, HG
	Cornwall	Cornwall Paper Mills Co.	N, C
	Fayetteville	McIntyre Paper	MP, N
	Fort Edward	Scott Paper Co.	PS
	Fulton	Armstrong World Industries	MP, N, C, PS
	Glens Falls	Finch, Pruyn & Co., Inc.	PS
	Greenwich	Stevens & Thompson Paper Co.	N, HG

State	City	Company Name	Type of Recovered Paper
	Hauppauge	Southern Container	C
	Hoosick Falls	Lydall, Inc.	C
	Little Falls	Burrows Paper Corp.	MP, PS
		Mohawk Valley Paper	PS
	Lockport	Domtar Gypsum	N, C
	Lyonsdale	Burrows Paper Corp.	PS
	Marcellus	Martisco Paper Co., Inc.	N, C
	Mechanicville	Tagsons Papers, Inc.	MP, N, C, PS, HG
	Newton Falls	Stora Papyrus Newton Falls, Inc.	PS, HG
	New Windsor	Lafayette Paper	MP, N, C
	Niagara Falls	Cascades Niagara	C
	Norfolk	McIntyre Paper Co.	PS
	Oakfield	USG Corp.	MP, N, C, PS
	Oswego	International Paper Co.	PS
	Plattsburgh	Georgia Pacific Corp.	HG
		Packaging Corp. of America	N, PS
	Portville	Fibercel Corp.	C
	Potsdam	Potsdam Paper Mills	PS
	Pulaski	Schoeller Tech Papers, Inc.	PS
	Rochester	Flower City Tissue	PS
	South Glens Falls	Encore Paper Co.	N, HG
	Walloomsac	Yorktowne Paper Mills, Inc.	MP, N, C, PS, HG
NC	Canton	Champion International Corp.	PS
	Charlotte	Caraustar Industries	MP, N, C, PS
	Cordova	Laurel Hill Paper Co.	HG
	Goldsboro	Celotex Corp.	MP
	Patersen	Sealed Air Corp.	PS
	Pisgah Forest	P. H. Glatfelter Co.	PS
	Plymouth	Weyerhaeuser	C, HG
	Riegelwood	Federal Paperboard	PS
	Roanoke Rapids	Champion International Corp.	MP
		Halifax Paperboard Co., Inc.	MP, N, C, PS
	Rockingham	Cascades Industries, Inc.	PS, HG
	Sylva	Jackson Paper Manufacturing Co.	C
OH	Avery	Certain-Teed Corp.	MP, C
	Baltimore	Fairfield Paper Co.	C
	Chagrin Falls	Ivex Corp.	MP, N, C, PS
	Chillicothe	Mead Corp. (Chilpaco)	PS
	Cincinnati	Cincinnati Paper Board Corp.	MP, N, C
		Rock-Tenn Co.	MP, N, C, HG
	Circleville	Jefferson Smurfit Corp.	C
	Coshocton	Stone Container Corp.	C
	Dayton	Fox River Paper Co.	PS
	Franklin	Georgia Pacific Corp.	MP, C

State	City	Company Name	Type of Recovered Paper
		Newark Boxboard Co.	MP, N, C, PS, HG
	Gypsum	USG Corp.	MP, N, C, PS
	Hamilton	Champion International Corp.	PS, HG
		International Paper Co.	PS
	Lancaster	Sonoco Products Co.	MP, N, C
	Lockland	American Tissue	MP, C, PS
		Jefferson Smurfit Corp.	MP, N, C, PS
	Massillon	Cleaners Hanger Co.	C
		Greif Board Corp.	MP, C
	Middletown	Bay West Paper Corp.	MP, N, C
		Crystal Tissue Co.	PS
		Jefferson Smurfit Corp.	MP, N, C, PS
		Newark Boxboard	MP, N, C, PS
		Sorg Paper Co.	PS
	Munroe Falls	Sonoco Products Co.	MP, N, C
	Rittman	Packaging Corp. of America	MP, N, C, PS, HG
	Toronto	Valley Converting Co.	MP, N, C, PS
	Urbana	Fox River Paper Co.	PS
	West Carrollton	Appleton Paper, Inc.	PS, HG
		Pentair, Inc. (Miami Paper Corp.)	PS, HG
OK	Ardmore	Georgia Pacific Corp.	MP, C
	Jenks	Kimberly-Clark Corp.	PS
	Muskogee	Fort Howard Corp.	MP, N, C, PS, HG
	Pryor	Georgia Pacific Corp.	MP, N, C, PS
		National Gypsum Co.	MP, N, C
		Orchids Paper (Robel Tissue Mills)	PS, HG
	Valliant	Weyerhaeuser Co.	C
OR	Albany	Willamette Industries, Inc.	C
	Clatskanie	James River Corp.	MP
	Corvallis	Western Pulp	C
	Gardiner	International Paper Co.	C
	Halsey	James River Corp.	HG
	Newberg	Smurfit Newsprint Corp.	N
	North Bend	Weyerhaeuser West Coast	MP, N, C
	Oregon City	Smurfit Newsprint Corp.	MP, N
	Pilot Rock	Wood Fiber Industries	N
	Springfield	Weyerhaeuser Co.	N, C, PS
	Toledo	Georgia Pacific Corp.	N, C
	West Linn	Simpson Paper Co.	PS, HG
PA	Chester	Scott Paper Co.	N, PS
	Downington	Brandywine Paperboard Mills	MP, N, C, PS
		Davey Co.	MP, PS
		Shyrock Brothers	MP
		Sonoco Products Co.	MP, N, C

State	City	Company Name	Type of Recovered Paper
	Erie	International Paper Co. (Hammermill)	PS
	Exton	Sealed Air (Exxon)	C
	Lancaster	American Paper Products Co.	MP, N, C
	Lebanon	Henry Molded Products	PS
	Lockhaven	International Paper Co. (Hammermill)	MP, PS, HG
	Mehoopany	Procter & Gamble Paper Products	PS
	Milton	National Gypsum Co.	MP, N, C
	Miquon	Simpson Paper Co	PS
	Modena	Sealed Air Corp.	MP, C
	Philadelphia	Connelly Containers	C
		Jefferson Smurfit Corp. (CCA)	MP, N, C, PS
		Newman & Co.	MP, N, C, PS
	Ransom	Pope & Talbot, Inc.	HG
	Reading	Caraustar Industries, Inc.	MP, N, C, PS
		Interstate Container Corp.	C
		Sealed Air Corp.	MP, C
	Spring Grove	P. H. Glatfelter	PS
	Stroudsburg	Rock-Tenn Co.	MP, N, C, ,PS
	Tyrone	Westvaco Corp.	HG
	York	Stone Container Corp.	MP, N, C, HG
		Yorktowne Paper Mills, Inc.	MP, N, C, PS
SC	Beech Island	Kimberly-Clark Corp.	PS
	Eastover	Union Camp Corp.	PS
	Florence	Stone Container Corp.	N, C
	Hartsville	Sonoco Products Co.	MP, N, C
	N. Charleston	Westvaco Corp.	PS
	Taylor	Caraustar Industries, Inc.	MP, N, C, PS
TN	Calhoun	Calhoun Newsprint Co.	MP, N
	Chattanooga	Caraustar Industries, Inc.	MP, N, C, PS
		Rock-Tenn Co.	MP, N, C, PS
	Counce	Packaging Corp. of America	C
	Covington	Lydall, Inc.	C, PS
	Harriman	Power Paper, Inc.	MP, C
	Kingsport	Mead Corp.	PS
	Knoxville	Tamko Asphalt Products, Inc.	MP, C
	Loudon	Kimberly-Clark Corp.	MP, PS, HG
	Memphis	Kimberly-Clark Corp.	PS
	New Johnsonville	Temple-Inland, Inc.	C
	Newport	Sonoco Products Co.	MP, N, C
TX	Daingerfield	Georgie-Pacific Corp.	MP, C
	Dallas	Rock-Tenn Co.	MP, N, C, PS
	Forney	Corrugated Services, Inc.	MP, C
	Galena Park	USG Corp.	MP, N, C
	Orange	Equitable Bag Co., Inc.	PS

State	City	Company Name	Type of Recovered Paper
		Temple-Inland, Inc.	C
	Pasadena	Simpson Paper Co.	PS
	San Antonio	Celotex Corp.	N, C, PS
	Sheldon	Champion International Corp.	MP, N
VT	Brattleboro	Specialty Paper Board, Inc.	MP, C, PS, HG
	East Ryegate	CPM, Inc.	N, C, PS
	Gilman	Simpson Paper Co.	PS
	Putney	Putney Paper Co., Inc.	HG
	Sheldon Springs	Rock-Tenn Co.	MP, N, C, PS
	St. Johnsbury	Elcon Corp.	PS
VA	Amherst	Virginia Fibre Corp.	C
	Ashland	Bear Island Paper	N
	Big Island	Georgia Pacific Corp.	C
	Buena Vista	Georgia Bonded	HG
	Covington	Westvaco Corp.	N, C, PS
	Edinburgh	Manville Corp.	N
	Franklin	Union Camp Corp.	PS
	Hopewell	Stone Container Corp.	C
	Lynchburg	Rock-Tenn Co.	MP, C
	Richmond	Custom Papers Group, Inc.	MP
		Manchester Board & Paper Co., Inc.	MP, N, C, HG
		Sonoco Products Co.	MP, N, C, PS
	West Point	Chesapeake Corp.	C
WA	Camas	James River Corp.	MP, N, HG
	Everett	Scott Paper Co.	MP, PS
	Longview	Longview Fibre Co.	MP, N, C
		North Pacific Paper Co.	MP, N
		Weyerhaeuser Co.	PS
	Millwood	Inland Empire Paper Co.	N
	Port Angeles	Daishowa America	MP
	Port Townsend	Port Townsend Paper Corp.	MP, N
	Sumner	Sonoco Products Co.	MP, N, C
	Tacoma	Jefferson Smurfit Corp.	MP, N, C
		Simpson Paper Co.	MP, N, PS
	Vancouver	Boise Cascade Corp.	HG
	Walulla	Boise Cascade Corp.	C
	Wenatchee	Keyes Fibre Co.	N
WV	Halltown	Halltown Paperboard	MP, N, C, PS, HG
	Wellsburg	Banner Fibreboard	MP, N, C, PS
WI	Appleton	Fox River Paper Co.	PS, HG
		Riverside Paper Corp.	N, PS
	Ashland	James River Corp.	PS, HG
	Beloit	Beloit Boxboard Co.	MP, C
	Brokaw	Wausau Paper	PS

State	City	Company Name	Type of Recovered Paper
	Cedarburg	Formart Containers	MP, N, C
	Cornell	Globe Building Materials, Inc.	N, C
	De Pere	U. S. Paper Mills Corp.	MP, C, PS
	Eau Claire	Pope & Talbot, Inc.	HG
	Germantown	Fiberform Containers, Inc.	N, C
	Green Bay	Fort Howard Paper Co.	MP, N, C, PS, HG
		Green Bay Packaging, Inc.	C
		James River Corp.	PS, HG
		Procter & Gamble Paper Products Co.	PS, HG
		Procter & Gamble Paper Products Co.	PS
	Kimberly	Repap Wisconsin	PS
	Ladysmith	Pope & Talbot, Inc.	HG
	Marinette	Scott Paper Co.	PS
	Menasha	Wisconsin Tissue Mills	MP, N, C, HG
		Mead Corp.	PS
		U. S. Paper Corp.	MP, N, C
		Whiting Paper Co.	MP, PS
	Merrill	International Paper Co.	PS, HG
	Milwaukee	Keiding, Inc.	MP, N, C
		Newark Boxboard	MP, N, C, PS
	Neenah	P. H. Glatfelter Co.	PS, HG
		Kimberly-Clark Corp.	PS
	Nekoosa	Georgia Pacific Corp.	PS
	Niagara	Niagara of Wisconsin Paper Corp.	PS
	Park Falls	Cross Pointe Paper Corp.	MP, PS, HG
	Peshtigo	Badger Paper Mills	HG
	Port Edwards	Georgia Pacific Corp.	PS
	Rhinelander	Rhinelander Paper Co.	MP
	Rothschild	Weyerhaeuser	MP
	Shawano	Little Rapids Corp.	PS, HG
	Tomahawk	Packaging Corp. of America	N, C
		Tissue Recycling Corp., Inc.	HG
	Whiting	Neenah Paper Co.	MP, PS
	Wisconsin Rapids	Consolidated Papers, Inc.	MP, N

APPENDIX C

The information in this section is a partial listing of the Directory of U.S. Recovered Paper Dealers as published in the October 1994 edition of PaperMatcher. It provides a state-by-state listing of dealers who buy, sell, and transport recovered paper and paperboard. These dealers often help their customers market their recovered paper.

For a copy of PaperMatcher with a complete listing of names, addresses, and telephone numbers, please contact the American Forest & Paper Association, Inc., 1111 19th Street, Suite 800, Washington, DC, 20036, (800) 878-8878.

State	City	Company Name
AL	Bessemer	Hatco, Inc.
	Birmingham	Dunn Business Service Co.
		Vulcan Recycling
	Greenville	PFJ Enterprises
	Mobile	Fiber Marketing, Inc.
		Ma Norden Co.
		Recycled Fibers of Alabama
AZ	Scottsdale	K C International, Inc.
CA	Aptos	Clarkehine Enterprise Co.
	Azusa	Omega Paper Co.
	Calexico	Chico Transfer & Recycling
	Chico	Lindon Trading Co.
	Compton	Los Altos Garbage Co.
	Cupertino	Western Pacific Pulp & Paper
	Downey	Weyerhaeuser Paper Co.
	Fremont	Progressive Paper Stock
	Fullerton	Bel-Art Paper Stock Co.
	Long Beach	Berg Mill Supply Co.
	Los Angeles	Cal-West By Products
		Les Mendelson & Associates, Inc.

State	City	Company Name
		Los Angeles Paper Box & Board
		West Coast Fibers
		Gilton Solid Waste Management
	Modesto	Consolidated Fibres-Settsu, Inc.
	Oakland	KMC Paper
		Secondary Resource Supply
		California Paper Stock
	Orange	Alameda Junk Co.
	Palmdale	V & C Co.
	Pasadena	Quality Paper Fibers
	Pico Rivera	LEF Trading Enterprises
	Pleasant Hill	Koply Sedae Corp.
	Pomona	Paper Scrap Specialist
		Smurfit Newsprint Corp.
		Unity Trading, Inc.
	Redwood City	Confidential Material Destruction
	Sacramento	K & R Paper Recycling
	San Bruno	A-1 Scrap Metal
	San Francisco	Berg Mill Supply Co.
		Fibre Source
		Fibre Trade, Inc.
		Weyerhaeuser Paper Co.
		Weyerhaeuser Paper Co.
	San Jose	Bonded Document Destruction
	Santa Ana	Commercial Waste Paper Co.
	South El Monte	Bestway Waste Paper Co.
	South Gate	L & J Reclamation Service, Inc.
	Torrance	Sutta Co.
	Ventura	Walker's Recycling, Inc.
		Northcoast Fibers
	Walnut Creek	Specialty Fibres, Inc.
		Fred's Trash Barrel
CO	Pueblo	Hellman Trading Co.
CT	Branford	M. Wilder & Son, Inc.
	Meriden	Rubino Brothers, Inc.
	Stamford	Willimantic Waste Paper Co.
	Willimantic	Fiber Brokers, Inc.
	Woodbridge	Florida Data Bank Group
FL	Auburndale	American Waste Paper Corp.
	Hialeah	All Florida Waste Paper Co.
	Miami	Interamerican Paper Corp.
		Morgan Price & Co.
		Omnisphere Corp.
		U. S. Fiber Export, Inc.

State	City	Company Name
		U. S. Fibre
		Wood Pulp & Paper Corp.
	North Palm Beach	Traders International Corp.
	St. Petersburg	ACC Recycling Corp.
GA	Atlanta	Atlanta Intercel
		Fulton Fibers
		O Z & Co.
		Vander Kley & Assoc. Ltd.
	Baxley	Recycling Equipment Service
	Canton	Interstate Paper Recycling
	Fayetteville	Recycled Fibers Southeastern
	Griffin	Southern Paper Brokers, Inc.
	Lilburn	Wilson Marketing Group
	Nashville	Jack Donaldson
	Norcross	Paper Recycling International
	Peachtree City	National Fiber Supply Co.
	Rome	Covenant Paper Stock
		Ira Levy & Assoc.
	Tucker	Southeastern Fibers, Inc.
IL	Addison	Fibers, Inc.
	Barrington	Paper Recovery
	Bensenville	Combined Resources, Inc.
	Charleston	Kurt Sandrisser Paper Stock Co.
	Chicago	Atlas Recycling, Inc.
		Avid Sales Corp.
		Herman Bailen
		Chicago Paperboard Corp.
		Columbia Paper Corp.
		Continental Paper Grading Co.
		Donco Paper Supply Co.
		Father & Son Salvage
		Great Lakes Secondary Fibres
		Howard Zuker Assoc.
		Intercel
		Leader Box Corp.
		Mid America Paper Recycling Co.
		Northwest Paper Co.
		S & H Fibre, Ltd.
		Skokie Valley Recycling
		Unique Salvage
		Western Fibers, Inc.
	Glen Ellyn	Wheaton Junk Service
	Maywood	Du Page Paper Stock Co.
	McHenry	Metropolitan Fiber

State	City	Company Name
	Mt. Vernon	Arlington Salvage
	Mundelein	Alan Josephsen Co.
	Naperville	Paper & Chemical Exchange
	North Chicago	C & M Recycling, Inc.
	Northbrook	Paper Chase Exchange
	Northlake	American Paper Recycling Corp.
	University Park	Recycled Paper Products
IN	Crown Point	Commercial Waste
	Indianapolis	Pioneer Fibers, Inc.
	South Bend	North Side Iron & Paper Co.
IA	Sioux City	Siouxland Recovery
KS	McPherson	Refuse Service, Inc.
KY	Covington	Corrcycle, Inc.
	Independence	Bavarian Trucking Co.
	Lexington	Harry Gordon Scrap Materials
	Louisville	Superior Fiber
LA	Covington	M. A. Norden Co.
	Hahnville	St. Charles Recycle & Retail
	Lake Charles	Recycling Services
	Monroe	Recycling Services
MD	Baltimore	ABC Box Co.
		Canusa Corp. Fibergroup
		Capitol Fiber, Inc.
		D. C. Intercel
		G & S Recycling Corp.
		Henry Blum, Inc.
	Catonsville	Hershman Recycling, Inc.
	Cheverly	World Recycling Co.
	Essex	West Street Industries
	Halethorpe	Vangel Paper, Inc.
	Joppa	RMSI Service, Inc.
		Wilmington Paper Co.
	Laurel	FLOM Corp.
	Monkton	Great Bay Paper Co.
	Rockville	Save the Forests Recycling
	Westminster	Petry's Junk Yard, Inc.
MA	Avon	S. Spiegel
	Billerica	Vel-a-Tran Paper Recycling, Inc.
	Boston	Schirmer Paper Corp.
	Chicopee	Willimansett Waste Co.
	East Longmeadow	Elm Fibers
	East Weymouth	North Shore Recycled Fibers
	Fall River	James C. Santos Trucking
	Hanover	Willaim Goodman Co.

State	City	Company Name
	Holyoke	Paper Mill Supply Co.
	Kingston	Capital Paper Recycling, Inc.
	Malden	North Shore Recycled Fibers
	Mansfield	American Paper Recycling Corp.
	Medford	Basic Paper Recycling
	New Bedford	A. W. Martin, Inc.
	North Adams	Shapiro & Sons, Inc.
	North Attleboro	Miller recycling Corp.
	Quincy	M. Sugarman & Co.
	Salem	North Shore Recycled Fibers
	Sharon	Hannah Paper Recycling Co.
	South Boston	A J & O Waste Co.
	Springfield	Harry Goodman, Inc.
	Stockbridge	Berkshire Salvage, Inc.
	Wilbraham	William Goodman & Sons
MI	Ann Arbor	T. Nalepka Waste Paper Co.
	Detroit	Condura Carton & Packaging
		Downtown Paper & Metal
		International Paper Recycling
	Grand Rapids	Environmental Cost Management, Inc.
		Krell Paper Stock Co.
	Gross Pointe	Aries Chemicals, Inc.
	Kalamazoo	Re-Bro, Inc.
		Thall Associates
	Monroe	F & F Specialties
		Southeastern Fibers, Inc.
	Newport	Eagle Transporters & Spotting
	Okemos	American Cellulose Sales
	Royal Oak	Royal Oak Waste Paper & Metal
	Waterford	Metro Fibres, Inc.
	West Bloomfield	Allan Blum Co.
	Ypsilanti	Action Disposal & Recycling
MN	Maple Plain	Kernic USA, Inc.
	St. Paul	Southeastern Fibers, Inc.
		Waldorf Corp.
MS	Natchez	Fiber Marketing International
MO	Bridgeton	Imperial Paper Stock Co.
	Joplin	Joplin Waste Paper
	Kansas City	Cook Paper Recycling Corp.
	St. Louis	Federal International, Inc.
NH	Plymouth	Kelley's Salvage
NJ	Belleville	Frank Neri Paper Recycling
	Berlin	Tab Shredding, Inc..
	Boonton	American Paper Co.

State	City	Company Name
	Carlstadt	Fred F.Carlo, Inc.
	Cherry Hill	Morris Fibres, Inc.
	Clifton	Frank Neri Paper Recycling
		Gaccione Brothers & Co.
	Closter	Miele Sanitation Co.
	Creskill	Abel Consultants Assoc.
		Tally Refuse Removal, Inc.
	Deptford	Gypsum Paper Fiber Co.
	Elizabeth	A & J Trading Corp.
	Elmwood Park	Garden State Paper Co.
	Fairfield	Recycling Management Systems
	Fort Lee	Prins Recycling Corp.
	Hillside	Thor Book Destruction
	Hoboken	Dell Recycling
		Hoboken Recycling Corp.
	Jersey City	A & J Trading Corp.
		A. Meluso & Sons, Inc.
		City Mill Supplies, Inc.
		Reliable Paper Recycling
		S. Morena & Sons, Inc.
		V. Ponte & Sons, Inc.
	Linden	A & J Trading Corp.
		A & J Trading Corp.-Industrial
	Morristown	American Paper Co.
	Newark	Industrial Refuse Removal
		Industrial Recycling
		Recycled Fibers International
		Regional Recycling Corp.
	Orange	Yan-Yan, Inc.
	Osbornsville	C & R Waste Materials Co.
	Paramus	Ocean Land Service
	Passaic	N & V Paper Recycling, Inc.
	Paterson	Annex Paperstock, Inc.
		R. Lobosco & Sons, Inc.
		Tabulating Card Salvage Co.
	Pennsauken	Daniel Parisi & Sons
	Piscataway	Waste Paper Buyers Co.
	Princeton	Princeton Waste Paper & Metal
	Ramsey	Valley Paper
	Rochelle Park	K-C International
	Secaucus	Allegro Sanitation Corp.
	Springfield	EST Fibers, Inc.
	Tenafly	Meeker Ave. Space Carting Corp.
	Wallington	Allied Waste Products, Inc.

State	City	Company Name
NY	Albany	Environmental Concepts
		Plesser Brothers Waste Paper
		Yank Waste Co.
	Amsterdam	Nathan's Waste & Paper Stock
	Binghamton	Indian Valley Industries, Inc.
	Bronx	Crystal Waste
		Crystal Waste Paper, Inc.
	Brooklyn	All City Paper Fibers Corp.
		Alpha Waste Paper Co.
		Barretti Carting, Inc.
		Chambers Paper Fibres, Corp.
		D. V. Carting, Inc.
		De Can Waste Paper Co.
		Victor De Vito
		G & B Carting Corp.
		Grand Waste Paper Corp.
		Highway Container Corp.
		James De Marco & Sons, Inc.
		Joe's Waste Paper Corp.
		John Rizzo & Sons, Inc.
		Lauro's Waste- Paper Corp.
		Liguori Paper Mills Supply Co.
		Litod Paper Stock Co.
		Lunati Brothers
		M & D Waste Materials, Inc.
		Main Paper Stock Co.
		Mele Paper Stock
		N. Sillaro & Sons
		Rapid Recycling Corp.
		Rinaldi Recycling Co.
		Rutigliano Paper Stock, Inc.
		Seco Waste Paper Co.
		Smith Waste Material Corp.
		Tocci Brothers, Inc.
		Williamsburg Paper Stock Co.
	Buffalo	Ramcol Fibres, Inc.
		Riverside Paper Recycling
		Tamco Paper Stock, Inc.
		Waste Stream Technology
		William Goodman & Son
		Wray Fibres International, Inc.
	Cheektowaga	Hannah Paper Recycling, Inc.
	Cohoes	Nathan H. Kelman, Inc.
	Coram	Brookhaven Recycling

State	City	Company Name
	Depew	Domtar Fiber Products
	Elmira	I. Shulman & Son, Inc.
	Flushing	Delmar Recycling Corp.
		NMS Waste Paper Removal, Inc.
	Fort Edward	North American Pulp & Paper
	Glens Falls	Perkins Recycling
	Hamburg	Canusa Corp.
	Jamaica	Epsee Trading Corp.
		Jamaica Paper Stock
	Kingston	Jordan Trading
	Lackawanna	Shred-It, Inc.
	Lake View	World Recycling Co.
	Latham	Ash Trading Corp.
	Mt. Vernon	Mt. Vernon Fibres, Corp.
	New York	Anchor Paper Stock Co.
		Apex Waste Paper Recycling Co.
		Atlas Paper Stock Co.
		D. Benedetto, Inc.
		Dell Recycling, Inc.
		F & N Wastepaper, Inc.
		Greater N. Y. Waste Paper Assn.
		M & G Carting
		Veterans Paper Stock & Mill Co.
		Vibro Carting, Inc.
	Newburgh	Formisano Recycling Center, Inc.
	North Bellmore	Monbro Sanitation Service, Inc.
	North Merrick	Liguori Carting Co.
	Port Chester	Paper Mill Supplies Ltd.
	Poughkeepsie	J C Paper Co.
	Rochester	Paper Mills Supply Co.
	Schenectady	Buff & Buff, Inc.
		T. A. Predel & Co.
	Staten Island	De Can Waste Paper Co.
		Stokes Waste Paper Co.
	West Babylon	All County Recycling, Inc.
	Westbury	Jamaica Ash & Rubbish Removal
	Williamsville	Great Niagara Paper
	Walloomsac	Yorktowne Paper Mills, Inc.
NC	Greensboro	AFM Stor
		Carolina Fibre Corp.
	Hickory	Paper Stock Dealers, Inc.
	High Point	Sealed Air Corp.
	Raleigh	ARH International
	Salisbury	Browning-Ferris Industries

State	City	Company Name
	Shelby	Cleveland Container Service, Inc.
		Paper Stock Dealers, Inc.
OH	Akron	WTE Corp.
	Canton	Marks Paper Stock Co.
	Cincinnati	Accu-Pak
		Cincinnati Paperboard Corp.
		Donco Paper Supply Co.
	Cleveland	Proshred Security
		Smith & Sons
		Sobel Salvage Co.
		Tavens Container Co.
	Columbus	Grossman Industries
		Royal Paper Stock Co.
	Dayton	Ira Levy & Assoc.
	Fairfield	Normac Fibers, Inc.
	Lima	Edith Rachlin
	Miamisburg	Dayton Fiber, Inc.
	North Canton	Sanford Maxson Co.
	North Lima	Associated Paper Stock, Inc.
	Norwalk	Norwalk Waste Materials Co.
	Van Wert	City Waste Paper
	Washington Court House	Cartwright Salvage Co.
	Willoughby	Willoughby Iron & Waste Co.
	Wooster	Metallics Recycling, Inc.
OR	Gresham	Arrow Sanitary Service
	Portland	Cascade Fibers International, Inc.
		K-C International Ltd.
PA	Bala Cynwyd	D K International Ltd.
	Bensalem	Viking Fibres, Inc.
	Blue Bell	Fibro Source USA, Inc.
	Braddock	Braddock Waste Materials Co.
	Clarks Summit	D K Trading
	Conshohocken	A. Masciantonio
		Vento, Inc.
	Coraopolis	Sarlo, Inc.
	Downington	Boiler House
	Eagleville	Gateway Disposal
	Easton	M S Reilly
	Erie	Automated Recycling Industries
	Harrisburg	Harrisburg Waste Paper Co.
		Spectrum Industries
	Huntingdon Valley	JIS Assoc.
	Levittown	Gabriele Paper Recycling, Inc.
	Lititz	RAM Corp.

State	City	Company Name
	McMurray	Ace Paper Recycling, Inc.
	Merion Station	Gaccione Bros. & Co.
	Morrisville	Gabrielle Paper Recycling, Inc.
	Narberth	Columbia Paper Corp.
	Norristown	Jenkins Paper Recycling
		Vento, Inc.
	Pennsburg	Lloyd Nace Waste Paper
	Philadelphia	American Disposal Systems
		Gilco Sales Co.
		Marchisello Brothers, Inc.
		Michael Ruzzo Industries, Inc.
		Pinto Brothers Recycling, Inc.
		V. I. Thomas Paper Co.
	Pittsburgh	ARCO Trading Co.
		G. H. Einhauser Co.
	Pottstown	Davis Brothers Scrap Co.
	Quakertown	Paper Chase Co.
	West Chester	First Fiber Corp. of America
	Wyncote	David A. Cutler Industries
RI	Pawtucket	United Box Co.
	Woonsocket	Woonsocket Waste Paper
SC	Cheraw	Plyler Paper Stock Co.
	Columbia	Confidential Control, Inc.
	Greenville	L & C Enterprises
	Hartsville	Sonoco Products Co.
TN	Germantown	Ira Levy & Assoc.
	Hendersonville	Retech GroupSouthern Bindery Service.
	Lebanon	JBC South, Inc.
	Memphis	American Fiber Manufacturing
TX	Austin	Weyerhaeuser Co.
	Carrollton	Material Resources, Inc.
	Cedar Hill	Allied Vista, Inc.
	Dallas	Cmacho Box Co.
		Pioneer Paper Stock Co.-Texas
		Pronapade Fibre, Inc.
		G. T. Technological Waste Control
	Dumas	Southwest Paper Stock Co.
	Fort Worth	Western Reclamation
		Weyerhaeuser Paper Co.
	Grand Prairie	AMP Metals & Paper
	Houston	Container Corp. of America
		Houston Advance Packaging Co.
		Oxbow Industries, Inc.
		Paper Recycling Service

State	City	Company Name
		World Fibers, Inc.
		Texas Shredding Co.
	Katy	Marco Export Ltd.
	Laredo	R. V. Wheelwright & Assoc.
	League City	Marolf Enterprises, Inc.
VT	Brattleboro	American Paper Recycling, Inc.
	South Burlington	Canusa Corp.
	St. Albans	South East Recycling Corp.
VA	Alexandria	Incendere, Inc.
	Chesapeake	Tri City Recycling Corp.
	Chester	Imperial Paper Stock Co.
	Richmond	Incendere, Inc.
		Recycling Services Co.
		Ree Tree Co.
		Pacific Coast Fibres, Inc.
WA	Kirkland	Waste Paper Services, Inc.
	Spokane	Smurfit Recycling Co.
	Vancouver	D & M Recycling
WV	Glen White	Quality Suppliers
	Piedmont	Channeled Resources, Inc.
WI	Appleton	Donco Paper Supply Co.
	Green Bay	Gleeson Fibers
		La Crosse Paper Recycling
	La Crosse	Milwaukee Waste Paper Co.
	Milwaukee	Peterman Paper Processing, Inc.
		Recycled Fibers Div.-Midwest
		Koplik Group
	Neenah	J & M Fibers, Inc.
	Sun Prairie	

GLOSSARY

Back-haul—Retrieval by a supplier or shipper of a shipping container or pallet from the delivery point.

Biodegradable—Material that micro-organisms can break down into basic chemical elements.

Biosphere—Living things and the environment of which they are an integral part.

Carcinogen—Substance that medical researchers have determined can cause cancer.

Chlorofluorocarbons (cfcs)—Chemicals that have been shown to destroy the Earth's protective ozone layer.

Closed loop—Using goods that have been remanufactured from recycled materials.

Cogeneration—Generation of energy by both the utility company and the consumer.

Commercial waste—Trash and/or garbage generated by businesses.

Commingle—Collect various recyclables together without separating them by type.

Compost—A soil conditioner, or humus, resulting from the breakdown of organic matter; not a fertilizer due to the high organic but low nitrogen content.

Contaminant—Item or material that damages or otherwise prevents an item from being recycled.

CPO—Computer print-out.

De-inking—The process of removing ink from recycled paper and newsprint.

Dioxins—Chemical byproducts primarily from manufacturing chemicals and from incomplete combustion of material containing chlorine atoms and organic compounds. Sources include MSW incineration, forest fires, auto exhaust, herbicides, and the combination of chlorine and wood pulp.

Divert—Avoid sending discards to the landfill by either reuse, recycling, or incineration.

Duplexing—Making double-sided copies.

Feedstock—Raw material for processing.

Generator—The person, business, or organization generating the waste.

Groundwood paper—Paper made from wood fiber by mechanical, not chemical, pulping processes.

Hauler—Agency, company or individual contracted to collect and transport waste (and recyclables) to a MRF, landfill, or transfer station.

Hazardous material/waste—Poisonous, corrosive, flammable, explosive, and/or radioactive substance that can endanger the health or well-being if transported, stored, or handled improperly.

Incineration—Disposing of waste by burning it under controlled conditions.

JRT—Jumbo roll (toilet) tissue.

237

Kenaf—Fast growing hibiscus cousin, producing 6-10 tons of raw fiber per acre with low lignin content. Requires fewer chemicals to convert to pulp than timber.

Landfill—Private or municipal site where solid waste is buried under layers of soil; the "dump."

Leachates—Liquid or gaseous byproducts of landfill materials as they decompose, such as methane.

MRF—Materials recovery facility where recyclables are separated. A "dirty MRF" is a facility where recyclables are recovered from commingled garbage.

Municipal solid waste, MSW—Total garbage and other discarded solid materials generated by households, commerce, and industry (as opposed to agricultural waste).

OCC—Old corrugated containers.

ONP—Old newspaper.

Pallet—Wooden platform used to move bales, large items, or boxes with a forklift; "skid."

Photodegradable—Material that will decompose when left exposed to light.

Pollutant—Material that is destructive or harmful to the environment.

Post-consumer—Materials that have served their initial use by the consumer.

Pre-consumer—Manufacturing waste that has not reached the consumer such as unsold newspapers, sawdust, and wood shavings.

RCRA—Resource Conservation and Recovery Act; federal law signed into effect October 21, 1976, for hazardous and solid waste management and procurement of materials from recovered wastes.

Reclaim—Recover and reuse materials.

Recovery rate—Percentage of usable recyclable materials removed from the waste stream.

Recyclable(s)—Good(s) that can be remanufactured, such as newspaper, aluminum, glass, and plastic.

Recycle—Collecting, separating, and processing pre- and post-consumer waste into new products. People often say "recycle" when they really mean "reuse."

Reduce—Consume and generate less waste.

Reuse—Using an item or material more than the intended number of times; may involve donating or reselling.

Secondary materials/fibers—Those that have been previously grown, mined, refined, or manufactured and used, and are to be used again to produce new goods.

Source reduction—Activities by the "source" (consumer, manufacturer, or producer) to reduce the consumption, amount, volume, and weight of products and packaging.

Source separation—Waste stream materials sorted by the generator, the "source," for collection.

Sustainable economy—An economy that prospers without harming the environment.

Tipping fees—Amount of money charged per ton for disposal at a landfill.

Toxic waste—Poisonous, dangerous to health or well-being if transported, stored, or handled improperly.

Transfer station—Intermediate facility to transfer waste from collection vehicles to bulk transport for delivery to the final disposal site.

Vermicomposting—Composting using special worms, usually red worms.

Virgin fiber—Fiber that comes from "natural" resources (forests, mines, and quarries).

Waste stream—Waste output of a community, facility, or region.

Waste to energy—Incineration or combustion of waste as fuel.

FOR FURTHER INFORMATION

Changes in addresses, area codes, and status of companies and organizations do occur. The author has made every attempt to verify the information herein prior to publication and asks the reader's understanding if any of the ones listed below may no longer be valid.

American Forest Council
1250 Conn. Ave., NW, #320
Washington, DC 20036
202-463-2455

Aluminum Assn.
900 19th St., NW
Washington, DC 20006

American Plastics Council
1275 K St. NW, #400
Washington, DC 20005
202-223-0125

ANPA
American Newspaper Publishers Assn.
Box 17407 Dulles Airport
Washington, DC 20041
703-648-1000

American Forest & Paper Assn.
1111 19th St., NW, #800
Washington, DC 20036
800-878-8878

Annenberg/CPB Collection
Race to Save the Planet Earth
South Burlington, VT
800-LEA-RNER

BioCycle
419 State Ave.
Emmaus, PA 18049
215-967-4135

CA Integrated Waste Management Board
8800 Cal Center Dr.
Sacramento, CA 95826
916-255-1000

CA Resource Recovery Assn.
4395 Gold Trail Way
Loomis, CA 95650
916-652-4450

Canon U.S.A., Inc.
Clean Earth Campaign
One Canon Plaza
Lake Success, NY 11042
800-962-2708

Danka - Environmental Business Solutions Recycle Center
2013-A Centimeter Circle
Austin, TX 78758
800-345-8898

EcoMedia Recycling Center
8016 Remmet Ave.
Canoga Park, CA 91304
800-359-4601

Environment Today and INFOLINK
1165 Northchase Pkwy., NE, #350
Marietta, GA 30067
404-988-9558

Environmental Defense Fund
1875 Connecticut Ave., NW
Washington, DC 20009
202-387-3500

Glass Packaging Institute
1801 K St., NW, #1105-L
Washington, DC 20006

Hewlett-Packard
11311 Chinden Boulevard
Boise, ID 83714
800-333-1917

INFORM, Inc.
100 Wall Street, 16th floor
New York, NY 10005
212-361-2400

Institute for Local Self-Reliance/ILSR
2425 18th St., NW
Washington, DC 20009
202-232-4108

Institute of Scrap Recycling Industries, Inc.
1325 G Street NW, #1000
Washington, DC 20005
202-737-1770

239

Investment Recovery Assn.
5818 Reeds Road
Mission, KS 66202-2740
913-262-4597

KP Products, Inc.
Kenaf Trailblazer Paper
P. O. Box 4795
Albuquerque, NM 87196
505-294-0293

Living Tree Paper Company
American-Milled Tree-Free
Paper
1430 Willamette St., #367
Eugene, OR 97401
503-342-2974

McRecycle USA
McDonald's Corporation
Kroc Dr., #062
Oak Brook, IL 60521
800-220-3809

National Consumers League
815 15th St., NW
Washington, DC 20006

National Materials Exchange
Network
509-466-1532

National Office Paper
Recycling Project
U.S. Conference of Mayors
1620 Eye St. NW, 4th Floor
Washington, DC 20006
202-223-3089

National Polystyrene
Recycling Company
4 Kildeer Court
P. O. Box 338
Bridgeport, CT 08014
609-467-9377

National Recycling Coalition
and Recycling Advisory
Council
1727 King St., #105
Alexandria, VA 22314-
2720
703-683-9025

National Solid Waste
Management Assn.
1730 Rhode Island NW,
#1000
Washington, DC 20036
202-659-4613

Newspaper Assn.
of America
11600 Sunrise Valley Dr.
Reston, VA 22091

Paper Recycler
600 Harrison St.
San Francisco, CA 94107
415-905-2200

Paper Stock News Report
Macantee Corp.
1327 Holland Rd.
Cleveland, OH 44142-3920
216-362-7979

Progress in Paper Recycling
and Directory of Recycled
Paper Mills
2323 East Capital Dr.
Appleton, WI 54915
414-832-9101

Recycled Products Business
Letter
Environmental Newsletters,
Inc.
11906 Paradise Lane
Herndon, VA 22071
703-758-8436

Recycled Paper Coalition
Community Environmental
Council
171 S. Fairfax Ave., #130
Los Angeles, CA 90036
213-933-6942

Recycling Manager
Capital Cities/ABC, Inc.
825 Seventh Ave.
New York, NY 10019
212-887-8528

Recycling Times
1730 Rhode Island Ave.,
NW, #100
Washington, DC 20036
202-861-0708

Recycling Today
P. O. Box 5817
Cleveland, OH 44101-9765
216-961-4130

Resource Recycling
P. O. Box 10540
Portland, OR 97210-1319
503-227-1319

Scrap Processing &
Recycling
1325 G. St., NW, #100
Washington, DC 20005

State Recycling Laws
Update
6429 Auburn Ave.
Riverdale, MD 20737-1614
301-345-4237

Technology Investment
Recovery , Inc.
2840 Broadway, Suite 331
New York, NY 10025
212-665-0184

UN Population Fund
220 East 42nd St.
New York, NY 10017
212-297-5011

Waste Age
4301 Conn. Ave., NW,
#300
Washington, DC 20008
202-244-4700

Worldwatch Institute
1776 Mass. Ave., NW
Washington, DC 20036
202-452-1999

INDEX

A

AB939 5
ACBC 69
Accountability 128
Africa 15
Air pollution 28, 31, 32
Albany 125
All Waste Paper Recycling, Inc. 71
Allen, Robert F. 52, 102, 167
Allenby, Brad 102, 103
Alonso, Ann Marie 132
Aluminum 60, 68, 194, 203, 205, 210
  can crusher 78
  can recycling bins 83
  cans 51, 58, 69, 76
  Cans for Burned Children 69
Amazon 15
America Online 180
American Forest & Paper Association 34, 101
Asia 15
Assessments 113, 194
Asset management 148
AT&T Recycles 77
Atlanta 88, 89, 95, 98, 170, 196
Audits 87, 100, 101, 132, 135, 169, 170, 182, 189, 194
Authority 174
Autistic Treatment Center 89
Award for Outstanding Achievement 101, 153

Awards 87, 88, 91, 101, 109, 110, 120, 122, 173, 190, 193, 195, 206

B

Babylon 39
Back-haul 185
Bag of the future 114
Balcones Recycling, Inc. 71
Ballwin 98
Barbados 6
Barium 202, 203
Basking Ridge 51, 152
Batteries 58, 92
Bauxite 17, 205
Bedminster 51, 53–58, 66, 67, 69, 94
Behind-the-counter 108, 126
Bell Labs 52, 63, 64
Benefits 16, 104, 109, 167, 176, 189, 191, 192
Benign 180, 188
  bleaching 121
Best Paper Recycling Award 101
Bethel New Life 95, 96, 97
Big Mac 116
Big Springs 79
Bins 24, 55–57, 60, 62, 67, 69, 70, 73, 77, 83, 84, 90, 91, 93, 123, 124, 134, 135, 138, 139, 143, 158, 185, 187, 190
Biodegradable materials 25, 67

Bleached 116, 121, 180, 188
  bags 116
  paper 120
Blue Angel 206
Blue Book 112
Bottom line 10, 25, 44, 75, 84, 126,
  144, 149, 154, 172, 174
Boxes 34, 53, 58, 59, 76, 89, 91,
  108, 114, 128, 153, 157, 181,
  185, 188
Boyhan, Walter 67
Brazil 15
Break-even analysis 203
Broadleaf forest 15
Broker 170
Brownstein, Mark 124, 175
BTUs 28, 45
Building Management Team 60
Building manager 125, 135, 136, 143,
  154
Building construction 11
Bulk 180, 201, 202, 204
Bulletin board 87, 179, 180, 189
Buy-back 96
Buying recycled 8, 32, 42, 43, 65, 66,
  86, 115, 117, 125, 145, 165, 187
Byron Western 142

C

Cadmium 92, 203
Cafeteria 40–42, 58, 59, 66, 67, 78,
  86, 87, 93, 103, 134, 145,
  169, 181, 187, 189, 190
Cairo Conference on Population 21
California 4–6, 14, 32, 33, 45, 100,
  124, 181
  Department of Conservation 181
Californians Against Waste Foundation
  32
Calling All Phone Books 112
Campaigns 27, 64, 69, 78, 81, 154,
  176, 189, 191, 193
Can crusher 78
Capacity 12, 13, 43, 81
Carbon
  dioxide 12, 16
  paper 59
Carbonless forms 58

Carcinogen 202
Cardboard 53, 55, 56, 63, 72–78,
  86, 89, 91, 93, 98, 100, 105,
  138, 157, 159, 188
Catalog 38, 88, 147, 155, 183
Cellophane 58
Central Hudson Gas & Electric 183
CENYC. *See* Council of the Environ-
  ment of New York City.
Chamber of commerce 174, 177
Champion Recycling Corporation 86
Champions of the Environment 102
Charities 60, 176, 186, 193, 209
Chemically processed paper 38
Chemicals 38, 202, 203
Chicago 94–97, 148, 155, 171
  Recycling Coalition 97
China 71
Chlorine 116, 120, 164, 180, 188
Chlorofluorocarbons (CFCs) 100–102,
  123
Clean-ups 154, 191, 194
Cleaning crew 58, 174, 176, 180
Clifton 157
Closed broadleaf forests 15
Closed Loop 25, 140–143, 145, 147
  Cooperative 140
  Recycling Conference 140
Closing the loop 8, 32, 42, 65, 85,
  115, 125, 141, 145–147, 155, 163
Coated paper 146
Cogeneration 210
Color-coded 124, 187
Colorado 137, 139
Colored
  dyes 6
  papers 58, 59
Columbia University 178, 179
Combustion 163
Commercial haulers 42
Commingling 152, 158
Committees 173–176, 189, 193–195
Community 174, 177, 191–193, 196,
  209
  Recycling 70, 95
Competition 55, 206
Competitive 8, 10, 44, 55, 102,
  127, 138, 154, 191

Competitor 121, 122, 124, 151, 153, 157, 174
Compliance 60, 61, 68, 83, 101, 128, 191, 192
Composting 34, 70, 114, 122, 123, 125, 128, 202
CompuServe 180
Computer 177, 178, 180, 182, 188, 189, 206, 210
  greenbar 121
  paper 151
  waste paper 132
Computerized 177
  companies 23, 133
Computers 23, 70, 71, 92, 120, 148
Connecticut 30
Conservation 109, 121, 122, 128, 140, 141, 181, 189, 194, 196
Consultants 112, 169, 170, 174–176, 186, 195
Consumption 9–11, 18, 21, 22, 28, 30, 31, 37, 44, 52, 57, 63–65, 69, 78, 80, 81, 105, 123, 128, 159, 163, 172, 174, 176, 195, 201, 203, 209
Containers 24, 41, 42, 67, 68, 72, 74, 76, 78, 87, 89, 90, 92, 93, 95, 103, 107, 114–116, 121, 126, 134–136, 138, 153, 158, 168, 170, 181, 184, 187, 204, 205
Contaminants 60, 61, 68, 83, 84, 87, 114, 143, 151, 153
Contamination 57, 67
Contract Services Organization 56
Cooperider, Allen 89
Corporate
  goals 104
  Information Technical Services 63
  Real Estate 131, 134, 139
Corrugated 42, 124, 125
  boxes 34, 53, 58, 108, 188
  cardboard 138, 157
Cost avoidance 98, 133, 134, 136, 137, 143, 148, 149, 154, 171–173
Council on the Environment of New York City 132, 134, 196

Cradle to reincarnation 104
Cradle-to-grave 100, 164, 206, 209
Crane's Company 142

D
Dallas 31, 71, 73, 75, 79, 88, 89, 91
Data Center 78–81
Deforestation 15, 16
DeNardo, Michael 137
Department of
  Consumer Affairs 133
  Environmental Protection 70, 153
  Sanitation 41, 174
Desert areas 15
Design For Environment 102, 104
Design for recyclability 164
Desktop
  recycling 137
  white 34, 38, 53, 54, 55
Developing nations 209
Dill, Michael 112
Dioxins 164
Direct Marketing Association 181, 182
Disposal
  costs 4, 6, 28, 44, 56, 93, 99, 151, 152, 168, 172, 174, 185, 204
  fees 32
  methods 4
Distributors 34
Diversion 5, 124
Document destruction 54, 153
Donations 90, 186, 204
Double-sided copies 44, 63, 64, 78, 178
Dry zones 15
Drycell batteries 58
Duke University 121
Dumpsters 60, 71, 169
Duplex copying 63, 105, 174, 178
Duplication 63
Dutchess County 5

E
Earnings 89
Earth Day 78, 86, 91, 92, 112, 140, 59, 190, 194
Earth Effort Packaging Awards 120
Easter Island 16, 17

EasyLink Services 99
Ebben, William 102
Eberhard 86
EcoMedia 204
Economics 109, 124, 148, 152, 159
EcoWriter 86
EDF. *See* Environmental Defense Fund.
Education, community 111
Educational materials 111, 128, 154
El Paso 79
Electricity 31, 32, 98, 120, 203
Electronic mail 23, 44, 59, 78, 105,
    172, 178, 180, 189, 195
Emissions 100–102, 202, 203
Employee
    awareness 120
    costs 30
    training 104
Endangered animals 92
Energy 9, 22, 28, 30–32, 38, 39,
    60, 69, 70, 102, 113, 120,
    122, 123, 128, 141, 146, 159,
    174, 175, 178, 181, 189, 203,
    206
    consumption 28, 31
    costs 69, 120
    sources 9
    Star 120, 189, 203
England 71
Environment and Safety Engineering
    105
Environmental
    and Conservation Award 109
    and Recycling Quality Team 98
    aspects of reuse 39
    Defense Fund 108, 109, 121
    degradation 16
    Excellence Award 87
    goals 57
    Health and Safety Department 76
    leadership 107, 109, 110
        Award 110
    messages 122
    mission 124
    Mug Program 59
    partnership 109
    protection 12, 47, 70, 104, 153,
        196, 201

    Agency 12, 24, 163
    policies 104
    Savings 75
    Vision Award 87
Environmentally
    friendly products 86, 87, 113
    preferable paper 121
Envision 86
EPA. *See* Environmental Protection
    Agency.
Erosion 121
European Union 5
Europeans 16
Executives 57, 175, 176, 193
Expenses 10, 32, 64, 66, 78, 81
Experts 77, 109, 128
Exports 6, 14, 15, 27, 28, 33, 34,
    209

F

Farmed trees 16
Feedstock 124
Fiberboard 202
Filters 120, 143, 180
Fingerhut 183
Florida 137, 139, 140, 159
Foam 73
Food waste 68, 70, 114, 124–126,
    137, 158, 169, 186
Foreign markets 33
Forest conservation 121
Forests 3, 14–17, 92, 128, 155
Formaldehyde 202
Fort Howard Paper Company 86, 151
Four Rs 176
Franchise 107, 124
Fuel 12, 16, 33, 39, 45, 67, 159,
    191, 203
Fuelwood 15
Fumes 159, 203
Furniture 3, 15, 69, 143, 155, 186,
    202, 204

G

Garbage 4, 5, 12, 14, 25, 28, 39,
    41, 42, 51, 56, 60, 61, 123,
    134, 136, 143, 152, 158, 159,
    164

Generators 43, 132, 137, 142, 159, 171, 182
Georgia 151
Georgia Pacific 122
Germany 12, 206
Gillette 201
Giordano, Anthony 54–57, 62, 63, 68, 71, 151, 152, 153, 155
Giordano Paper Recycling Corporation 54, 55, 151, 153, 177
Glass 5, 33, 41, 42, 51, 58, 60, 68 78, 92, 93, 103, 136, 138, 155, 159
Global
  environment 91
  Information Systems 52
  Real Estate 52, 56, 63
  roundwood production 15
Glossy paper 59, 203
Glues 164, 202
Goals 57, 94, 99, 101, 102, 104, 105, 157, 158, 163, 168, 171, 173–176, 191, 193, 195
Goodwill 196
Government 104, 142
  agencies 5, 8, 96, 117, 142
  decision makers 141
  officials 96
  regulations 5, 104
Grades 154, 159, 171
  fiber 43
  newsprint 37
  paper 33, 35, 53, 54, 76, 89, 133, 136, 139, 151, 154
Graedel, Tom 102
Great Depression 44, 136
Green Bay 151
Greenbar 34, 54, 76, 89, 121, 132, 158, 188
Gridcore Systems International 202
Groundwood 38, 43, 158
Guidelines 10, 113, 135, 172

H

Hampers 40, 57, 62
Handbook 57–59, 194
Happy Meal 116, 128
Hardwood 15

Harmful toxins 39
Haulers 5, 40, 42, 43, 84, 112, 124, 133, 137, 139, 141, 145, 146, 157, 169
Hauling 54, 56, 72, 74, 75, 98, 132, 133, 139, 143
  costs 75
Hazardous waste 77, 87, 92, 109
Health 102, 196
  expenses 32
  hazard 67
Heart Song 183
Hitachi 206
Home Box Office 141
Houston 86
Hub concept 89, 95, 98, 196
Human resources 83, 144, 179, 190
Humus 125

I

Illinois Secretary of State 183
Illnesses 16
Image 173, 177, 193
  public 117
Immigration 22
Incentive DP 85
Incentives 57, 64, 125, 181, 183, 190
Incineration 32, 39, 121, 159, 203
Incinerators 39, 43, 67, 159
Income 33, 89, 136, 176
Indonesia 15
Industrial
  ecology 102
  nations 209
  papers 11
  parks 168
  production 32
  purposes 15
Industrialized countries 12, 22, 36
Information
  age 21
  protection 55
  superhighway 44, 99, 180
Inks 6, 164, 202
Innovation 101, 122, 123, 125
Institute for Local Self-Reliance 96

Integrated Waste Management
  Act 5
  Board 33
International
  agreements 33
  markets 34
  Paper 142
Internet 180
Investment Recovery Association 148
Investors 96, 192

J

Jacksonville 91
Janitors 60–62, 70, 76, 86, 94, 104,
  133–135, 137–139, 143, 146,
  174, 176
Japan 18, 169
Jars 60
Jermyn, Steve 116
Jersey City 136, 137
Jobs 27, 32, 33, 60, 63, 84, 89,
  96, 125, 137, 141, 146, 151,
  158, 179, 190, 205
Johnson & Johnson 121
Jumbo rolls 59
Junk mail 24, 54, 77, 138, 152,
  181, 182

K

Kansas City 99
  Service Center 99
Kenaf 184
Kill-A-Watt 69, 70
Kinney Corporation 178
Kraft, Charlotte 167
Kroc, Ray 110
Krupp, Fred 109

L

Labor 30, 61, 63, 70, 76, 135
  costs 30, 76, 151
Landfills 4–6, 12–15, 24, 25, 27, 28,
  30, 32, 41, 42, 57, 59, 67, 68,
  72–74, 78, 93, 98, 99, 104, 112,
  115, 121, 123, 124, 142, 148,
  152, 159, 163, 184, 185
Langert, Bob 109, 110, 114, 116,
  122, 123, 126, 127, 175

LaPerna, Cheryl 54, 65, 175
Laser printers 34, 51, 59, 66, 77,
  81, 85, 178
Laughlin, Stephen 141, 142
Lauro, Andy 131, 134, 143, 146
Laws 5, 6, 41, 70, 121, 133, 137,
  144, 169
Leadership 70, 94, 98, 101, 104,
  105, 107, 109, 110, 120, 121,
  124, 128, 142, 144, 176, 193
League of Women Voters Population
  Coalition 5
Legislation 5, 40–42, 109, 112, 121,
  140, 151, 154, 163, 164, 206
Legislative strategies 96
Life-cycle assessment 113
Lion's Club 92
Lippe, Pamela 140
Liquid Paper 201
Litter 68, 70, 110
Livingston 69
Local Law #19 5
Logging 15, 17
Lombard, Phil 78, 175
Los Angeles 28, 31, 124, 155
Lumber 3, 15, 16, 155, 204

M

Madagascar 15
Magazines 30, 34, 38, 42, 53, 58,
  60, 86, 93, 135, 152, 164
Magnetic media 204
Magnuson, Keith 109
Mail Preference Service 182
Mailings 182, 183
Maintenance 32, 63, 83, 146, 154
  costs 63, 154
Makower, Joel 123
Management support 131
Managerial transformation process 167
Mandates 5, 41, 58, 76, 133, 144,
  147, 154
Mandatory
  recycling 133, 135
  source separation 41
Manhattan 40, 134, 185
Manufacturing process 34, 38, 100,
  101, 104, 188, 189, 202

paper 6, 34
  waste 100
Mardi Gras 86
Market share 174, 176
Markets 34, 43, 86, 117, 140, 141
Marriott 67
Mass burn 39
Massachusetts 39
Materials recovery 40
Matsushita 206
Maximizing 167
  capabilities 23
  investments 148
  revenues 136
Maximum 133, 204
  rate 149
  use 128
May, Marilyn 71, 106, 175
McCauley, Ron 81, 175
McDonald's Corporation 107, 112,
  123, 126, 184
McRecycle USA 115, 117, 120,
  122, 125, 126, 128, 174, 176
McWorld Stamp Design Contest 112
Measurements 128, 177, 194
Merrill Lynch 25, 131–134, 137–148
Methane 67
Mexico 4, 71
Midwest 43
Miller Business Products 85
Mills 36, 38, 43, 55, 71, 121, 138,
  141, 145, 146, 153, 154, 157,
  159, 164, 209
Milton Bradley 157
Minimization 44, 57, 59, 78, 100,
  174, 195
Minimizing 44, 136, 158, 167
Minimum 6, 85, 116, 128
Mission 109, 124, 189
Mixed
  loads 152
  office waste 34
  paper 40, 54, 74, 76, 89, 93,
  134, 138, 139, 152, 153, 154,
  158, 186
Model Safety Program 102
Monitoring 57, 79, 101, 104, 128,
  135, 154, 176, 189, 194

Montana 15
Monthly Cost Savings Worksheet 172,
  194
Morale 10, 191
Motivation 77, 94, 135, 176
Motor oil 60
MRFs 40, 42, 96, 159
MSW levels 4, 18
Multi-tenant buildings 41, 154, 170
Municipal solid waste 4–6, 12, 13,
  27, 28, 97

N

NAFTA 33
Napkins 86, 112, 114, 116, 128,
  145, 169, 180, 187
National Office Paper Recycling
  Challenge 173
National Office Supply 147
National Paper Recycling
  Coalition 136
  Project 87
Nations Bank 121
Natural
  fiber 11, 25
  resources 3, 10, 16, 22, 29, 44,
  113, 128, 145, 146
NER Data Products 148
Netherlands 125
New Jersey 25, 51, 53–56, 58, 69,
  70, 94, 134–139, 151, 153, 157,
  175, 178
New York 5, 28, 32, 39, 40, 55,
  79, 131, 132, 133, 135–137,
  140, 142–144, 153, 157, 165,
  183, 184, 196
Newark 151
Newsletters 60, 92, 97, 136, 139,
  171, 188
Newspaper 30, 34, 38, 39, 53, 58,
  60, 78, 86, 93, 120, 122, 135,
  138, 143, 193
  fiber 86
Newsprint 11, 36, 37, 42, 54, 86,
  116, 122, 152, 153, 158
  trayliners 122
Niagara Falls 31
Nickel cadmium batteries 92

Non-biodegradable food containers 68
Non-recyclables 42
Noncompliance 60
Nonprofit organizations 8, 89, 95, 97, 143, 204
Nonrenewable materials 164
North Carolina 155
Nuclear power plants 210
NYC Department of Sanitation 41

O

Oak Brook 114
Objectives 167, 172–174, 177, 191, 194, 195
Obstacles 60, 70, 83, 84, 190
Occupancy costs 132
OECD 14
Office
  buildings 40, 42, 125, 157
  centers 168
  paper 34, 38, 43, 44, 54, 58, 116, 151–153, 173, 177, 188
  services 63, 68
  supplies 59, 68, 85, 143, 201
Ohio 101, 188
Oil 17, 31, 32, 57, 60, 91, 98, 136
Old growth trees 16, 155
On-line 44, 79, 83, 159, 172, 178, 209
On-screen 44, 178
One-can system 158, 186
Operating costs 81, 105
Optimal Packaging Team 112
Ordinances 94, 97
Organic wastes 123, 125, 202
Organizations 121, 125, 167, 177, 191, 193, 204
Orientation 57, 83, 144, 190, 194
Outstanding Achievement in Recycling 70, 153
Over-the-counter waste 108
Oxygen 3, 14, 120, 180, 188
Ozone 92, 100, 178

P

Pacific Bell SMART Yellow Pages 112

Packaging 3, 5, 11, 12, 24, 59, 60, 85, 86, 92, 107–110, 112–117, 120–123, 126–128, 163, 164, 177, 180, 181, 184, 185, 188, 201, 202
Paine Webber 143
Pallets 15, 51, 72, 73, 114, 185, 186, 204
Paper
  bag 116
  consumption 10, 21, 30, 37, 44, 52, 63, 80
  costs 63, 66
  products 10, 12, 16, 27, 32, 34–36, 39, 43, 53, 65, 85, 86, 121, 145, 146, 152
  stock 147
  Task Force 121
  towels 24, 86, 145
Paperboard 10–12, 27, 34, 36, 157
Paperless
  office 23, 78
  society 23, 44
Patrons 126, 181
Payroll 179
Peanuts 73, 90, 185
Penalties 169, 191
Pennsylvania 17, 55, 153
Performance review 57, 60, 176
Perry, George 64, 175
Perseco 123
Personnel 58, 73, 78, 143, 173, 179, 202
Philosophy 67, 104, 107, 144, 175, 189, 192, 193, 194
Phone books 42
Phosphorous 43
Photocopier 24
Planet Protectors 76, 77, 83–89, 93, 94
Plants 16, 28, 34, 39, 86, 96, 142, 159, 184, 210
Plastic 3, 5, 27, 41, 42, 59, 60, 62, 67–70, 73, 76, 78, 89, 91–98, 114, 117, 122, 136, 143, 158, 159, 169, 181, 184, 186, 204–206, 210

Plates 67, 69, 181, 187, 203
Playland 115, 116
Plywood 202
Policies 57–60, 63, 64, 101–104,
   127, 128, 136, 140, 172, 194
Pollution 14, 28, 31, 32, 39, 60, 101,
   104, 116, 120, 122, 164, 188
Polynesians 16
Polystyrene 66–68, 73, 76, 90, 93,
   107, 110, 121, 125, 126, 158,
   159, 185, 205
Population 3, 5, 6, 14, 16, 21, 22,
   28, 80, 91, 105, 163, 194
   Week 194
Portland Recycling Congress 114
Post-consumer
   content 6, 116, 169, 187
   fiber 116
   material 6, 42, 190
Post-industrial materials 129, 163,
   187
Postage 112, 180, 182
Posters 64, 187, 191, 192
Power Systems Division 71, 75, 88,
   175, 176, 178–190
Pre-consumer
   content 116
   fiber 34
Preferred packaging guidelines 113
Print Reduction Committee 144
Printed reports 178
Printers 34, 38, 42, 51, 59, 65, 66,
   77, 81, 85, 178, 179, 188, 189
Printing 6, 11, 34, 43, 44, 59, 78,
   81, 99, 121, 137, 139, 140,
   144–147
   costs 81
   paper 43, 121, 145
Process modeling 100
Production costs 65
Profits 10, 23, 45, 113, 155, 167,
   193
Property manager 54–57, 63, 64, 70,
   103, 131
Proprietary documents 58, 59, 71
Prudential 121
Prudential-Bache 143

Public relations 60, 64, 89, 136,
   141, 182
Publications 30, 144, 145, 171, 179,
   184
Publicity 77, 174
Pulp 31, 38, 201
Pulping 38, 152
Punch cards 131

Q
Quality
   improvement 72, 76
   policy deployment 101, 102
   team 76, 83, 98
Queens 143
Quick-copy 63, 64

R
Rainforest Action Network 17
Rainforests 15–17, 92, 128
Re-engineering 172, 209
Re-tree-bution 183
Receptacles 54, 58, 68, 104, 126,
   136, 186, 190
Records Management 56
Recovered paper 33
Recovery rate 12, 27, 38, 55, 57,
   124, 195
Recycled
   content 5, 6, 9, 10, 36, 43,
   44, 53, 116, 122, 125, 141,
   169, 177, 183, 184, 187
   fiber 6, 25, 29, 32, 43, 65,
   85, 116, 141, 145, 146, 154,
   155, 159, 187, 188
   products 8, 25, 42, 65, 86, 112,
   115–117, 125, 140, 141, 147
   Relay DP 142
   stock 65, 66, 86, 122, 146, 147
Recyclers 32–34, 38, 54, 55, 60,
   71–74, 84, 92, 98, 106, 141,
   146, 153, 170, 176, 186, 206
Recycling
   Advisory Council 6
   centers 96, 121
   coordinators 56, 76, 77, 83, 87,
   174, 175, 191

Council of Dallas 88
Hotline 139
programs 24, 158, 165, 186, 203
Trophy 88
revenues 132
Team 54, 55
Redesign 164, 176, 188–190, 206
Reduce, reuse, and recycle 6, 123
Refabricate 138
Regulations 5, 8, 41, 95, 104, 132, 169, 173, 196
Reimbursement 25, 38, 43, 54, 57, 72, 74–76, 86, 98, 99, 115, 132, 139, 143, 153, 154, 194
rates 25, 38, 43, 72, 74, 75, 86, 98, 115, 139, 143, 153, 154, 171, 186
Renewable materials 126
Rensi, Ed 107, 115
Report Management Documentation System 79
Reports 23, 57, 77–79, 81, 84, 94, 120, 131, 136, 139, 144
Reproduction 63
Reprographics services 63, 64, 66
Requests for proposal 142
Resource
Center 95
Conservation and Recovery Act 47
Retailers 34, 113, 181
Return on investment 168, 175, 203
ReUsables 185
Revenue 10, 24, 33, 39, 44, 51, 70, 72, 74–77, 94, 98, 132, 133, 136, 139, 149, 152, 195
Rewards 89, 177, 191, 192
RFP. *See* Request for proposals.
Rhode Island 30
Roguish, Fred 183
Roundwood 15, 16, 158

S
San Diego 112
San Francisco 46, 155
Schools 191, 209
Scrap 34, 37, 51, 55, 58, 69, 73, 83, 96, 97, 153, 204, 205
Scrubbers 39

Secondary fiber 43, 53, 151
Self-service copier 63, 64
Semass 39
Seventh Generation 183
Shredded 57, 59, 71, 74
paper 62, 71
Shredders 57, 71, 153, 154, 185
Shredding 57, 59, 71, 139
Siemens Nixdorf 206
Single-sided copies 24, 44, 63, 64
Smog 100, 203
Soil 16, 17, 27, 121
Solar panels 210
Solutions for the Earth 91
Solvents 100, 101
Sorgen, Howard 144
Source
reduction 9, 39, 43, 44, 78–80, 97, 108, 113, 114, 117, 125, 163, 164
separation 5, 41, 170
South Pacific 16
Southeast Asia 15
Specifications 177
Springhill 85, 142
St. Barnabus 69, 138
Stinson 51
Stone Container 116
Strategies 5, 9, 79, 96, 104, 105, 163, 165, 167, 174, 177, 190, 192–194
Subscriptions 30, 179
Suppliers 65, 69, 85–87, 107, 113, 114, 120, 122, 128, 147, 153, 174, 176, 177, 180, 186, 187, 202, 203
conference 120
Surplus stock 59
Sustainable
development 164
economy 140

T
Tactics 105, 165, 167, 177, 191
Taiwan 71
Task force 109, 112, 121, 123, 152, 174, 175, 189, 195
Teams 100

Teamwork 120, 144, 192
Technical Support Organization 64
Technology 10, 23, 43, 44, 65, 86, 102, 125, 127, 133, 146, 148, 154, 164, 172, 197, 209, 210
Telephone books 77, 93, 179
Tenant 41, 125, 154
Texas Recycles Day 90
3M 85
Throw-away 6, 23, 204
Ticonderoga 142
Timber 15, 16, 114, 169, 186
Time line 173, 177
Time Warner 141
Tipping fees 14
Tires 60, 116, 155
Tissues 11, 169
Toner 66, 77, 85, 148, 179, 201
Top management 167, 174–177
Topsoil 121
Total Quality Management 78, 104
Total recycled fiber 6
Towels 24, 86, 145, 146, 169, 180, 187
Toxic 159, 163, 201, 202
  air emissions 101
  solvents 101
  waste 77, 121
Toxins 39, 159, 180, 202, 206
Toys for Tots 89, 90
TQM. *See* Total Quality Management.
Trade deficit 33
Trailer 73
Training 23, 78, 83, 94, 97, 104, 172, 195
Trammel 159
Trash compactor 56
Trash on a Rope 94
Trayliners 122
Tree farms 16, 25, 34, 155
Tree-based paper 184
Treeless wasteland 30
Tritech 137, 139, 146
Tropical hardwoods 15
Tully, Daniel P. 144, 148
Tuscaloosa 86
Twardy, Jerry 54, 56, 58, 175
Two-can system 55–57, 138, 153

U
U.S.
  EPA 6, 187
  foreign trade deficit 33
  Postal Service 112
Unbleached 116, 180, 188
  paper 120
  recycled carry-out bags 116
Under-the-desk recycle cans 58, 68, 94, 103, 152, 190
Unity DP 85
Unity Environmental 157, 159
Unshredded 139
Unsustainable 4
Upper management 54, 70, 72, 105
UV 100

V
V. Ponte & Sons 157, 158
Van Orden, Jim 89
Vendors 55, 62, 68, 73, 76, 85, 86, 92, 113, 120, 127, 146, 154, 176, 177, 184–186, 188, 195, 196
Vermicomposting 123
Vickers, Richard 72, 175
Vincent, Keith 93
Virgin 65, 146, 147
  fiber 25, 28, 31, 36, 37, 43, 86, 155, 172, 183, 188
  groundwood 38
  material 164, 169
  paper 66
  products 65
  pulp 31
  stock 146, 147
Voice mail 179
Volunteers 92, 175, 176, 192

W
Warrenville 98
Washington Business Recycling Award 101
Washington, DC 30, 112, 136
Waste
  disposal 4, 10, 28, 33, 44, 53, 83, 93, 94, 99, 100, 102, 125, 151, 168, 171, 172
  generation 14

generators 137, 142
haulers 42, 124
management 5, 33, 39, 54, 56, 70, 95, 102, 107, 109, 124
minimization 59, 174, 195
paper 6, 29–31, 44, 55, 63, 64, 71, 86, 88, 89, 91, 94, 98, 132, 134, 135, 138, 139, 141, 143, 145, 152, 153
Prevention & Recycling Service 196
profile 169
reduction 5–7, 33, 60, 71, 83, 94, 95, 107, 109, 110, 113, 114, 116–118, 127, 128, 142, 157, 158, 163, 173, 176, 190, 194
  Action Plan 107, 113, 114, 116–118
  goal worksheet 173, 194
  Packaging Specifications 113
  program 116
stream 5, 12, 40, 59, 98, 100, 107, 108, 124, 127, 136, 164, 165, 169, 183, 201, 204
Watcher 99

Waste-to-energy 28, 39, 159
Wasteland 16, 30
Wattenizer 69
We Can 142
Weight of paper used (WPU) 183
Westwood 98
  Ecology Quality Circle 98
Wet
  rubbish 137
  waste 137, 143
White
  ledger 42
  paper 58, 59, 132
Wood 11, 15, 16, 28, 38, 43, 69, 72, 73, 76, 114, 117, 155, 158, 186, 202
Workshop 96, 140, 189
World Financial Center 131, 133
World Trade Towers 79

X

Xerox 81

Y

Yastrow, Shelby 121, 175